Climate and Energy Governance for the UK Low Carbon Transition

Thomas L. Muinzer

Climate and Energy Governance for the UK Low Carbon Transition

The Climate Change Act 2008

palgrave
macmillan

Thomas L. Muinzer
Stirling Law School
University of Stirling
Stirling, UK

ISBN 978-3-319-94669-6   ISBN 978-3-319-94670-2   (eBook)
https://doi.org/10.1007/978-3-319-94670-2

Library of Congress Control Number: 2018948722

This Palgrave Pivot imprint is published by the registered company Springer Nature Switzerland AG
The registered company address is: Gewerbestrasse 11, 6330 Cham, Switzerland

Foreword

Holland is the latest in a long list of countries that are copying the UK's Climate Change Act. From Mexico to Sweden to New Zealand, our legislation has been hailed as a game-changing example. This book is therefore particularly welcome and timely. Dr. Muinzer is the first to assess the contribution of the Climate Change Committee which was born of that Act a decade ago, and his account is an important contribution to our debates about our second ten years.

Back in 2008, the Labour government was in power when an alliance between the Conservatives and Friends of the Earth produced a draft Act. Peter Ainsworth was the particularly far-sighted Shadow Environment Secretary who had been encouraged by the new Tory Leader, David Cameron, to push out the boundaries of the Party's policy and embrace the green campaigners. Thus, the Climate Change Act was born and all the opposition parties signed up to it—Liberal Democrats, Nationalists and independents, with the support of over 100 Labour MPs who called on the Government to support the proposal.

Ed Miliband, Secretary of State for Climate Change, was a supporter instinctively but the Treasury, and therefore the Prime Minister, was very wary that the new Act would restrict their freedom of action. It wasn't that they didn't accept the need to combat climate change, but they hated the idea that there would be what they saw as a competing series of budgets—statutory carbon budgets. It's been the same in every other country since. Governments balk at the idea of a truly independent body setting carbon budgets, holding them to account and being supported by the full force of law.

Yet this is the genius of the Act. It recognises that we have to reconcile two timelines: the democratic need for regular refreshment of mandates through elections and the absolute need for a continued and long-lasting policy base to fight climate change. The first is five years or less. The second lasts at least until 2050. The reconciliation offered by the Climate Change Act is that the committee of entirely independent scientists and economists recommend the budgets well ahead. Parliament votes on them. They are then statutory requirements and cannot be changed except on the recommendation of the Committee.

So it is that we now have budgets till 2032, and we shall start on 2032–2038 this coming Autumn. But we have much more to do. We must involve an ever-wider number of our people and organisations, local authorities, voluntary groups and Churches in working to reach our goals. We need to help the government realise its commitments in Paris. We need to instil a much greater sense of urgency—particularly in the adaptation agenda. Above all, we must not let up in our efforts at both home and abroad.

Britain was a creative pioneer in passing the Climate Change Act. It has done spectacularly well in decarbonising its power system. It now needs to show the same commitment in transport and agriculture. We are in the strongest position to make it clear how the battle is to be won and inspire others round the world to step up to the mark. It is our institutional innovation that makes it possible. It's the Climate Change Act that keeps our feet to the fire.

Suffolk, UK   The Rt Hon. Lord Deben

Preface

The UK's Climate Change Act provided the first instance in the world where a country placed blanket legally binding long-term emissions reduction targets upon itself in order to combat climate change. As such, in this crucial area of environmental governance, UK law and policy has pioneered a very particular form of national framework, and many other countries are beginning to follow suit, with some modelling their approach, or aspects of their approach, on the UK method. The governance of climate change is a major issue of our times; in producing this study, my intention has been to provide the first book-length elaboration and analysis of the important UK Climate Change Act framework, written in a concise and to-the-point manner commensurate with Palgrave's *Pivot* series format.

While one can read or dip into this book for general scholarly or intellectual interest, it is also intended to fulfil a practical function, that is, to enrich the presently limited pool of knowledge concerning the UK's pioneering climate framework, and in doing so to provide the only comprehensive introduction to the UK's primary climate Act. The central intention has been to accessibly introduce, explain and explore this complex climate regime, doing so in a clear and interdisciplinary manner that is designed to open its workings up to a broad audience, i.e. an audience that is not rooted narrowly in the tradition of technical environmental law, bearing in mind that the Act itself is a complex legislative framework.

This study provides: coverage of the background to the framework and its importance to the UK Low Carbon Transition, including the political 'story' around the lead-up to its creation; explication of the technical content of the Act, and consideration of its practical application; recognition, elaboration and exploration of key thinking and debates surrounding the framework in the scholarly environmental literature; exploration of the Act's international significance, and its internal 'subnational' dimensions and impact, engaging the UK's devolved territories of Northern Ireland, Scotland and Wales.

The UK's climate framework engages all of the UK's major socio-economic sectors; however, in this study the 'energy sector'—treated here as encompassing electricity and heat generation—has been chosen to occupy the central sectoral focus of analysis due to the fact that energy generation amounts to the single largest source of greenhouse gas emissions in the UK (and across the international community as a whole). As such, it is the sector that has been most directly and pervasively acted on by the 2008 framework to date, and the high level of emissions arising from energy means that adequate energy decarbonisation is essential to the UK Low Carbon Transition's success. Further, if a book of this nature attempted to focus in detail on the UK's socio-economic sectors 'in general', it would spread itself very thinly indeed.

Given both the introductory and analytical nature of this study, acting in combination with the broad reach of the matters engaged and impacted by the Act itself, this book endeavours to provide a first, much-needed interdisciplinary treatment of the framework that should hold appeal and be useful to:

- Interdisciplinary university audiences at undergraduate and postgraduate level;
- Climate change policy practitioners and the wider policy community, including in the UK, other European countries, the EU Commission and potentially those in government and regulatory bodies beyond Europe;
- Academics working within environmental studies, environmental policy and governance, and environmental and energy law (UK, EU and international specialists);
- Think tanks nationally and internationally;
- NGOs and philanthropy;
- General interest readers seeking an introduction to the field;
- Specialised media analysts;
- Politicians being lobbied to adopt this form of legislation.

The governance regime sketched out under the terms of the Climate Change Act may be far from perfect. Nevertheless, the UK did do something of undeniable importance in 2008 when it put in place the first national Act of Parliament in the world that saw one nation place long-term legally bindinggreenhouse gas emissions reduction targets on itself in legislation, and the Act also included a range of other pioneering climate-related legal duties, extending to climate adaption in addition to climate mitigation. This important Act is firmly on the statute books in the UK at the time of writing, that is to say, it is still active law, and based on current trends, it is likely to remain in place for a long time to come. As such, it is fitting that this edition has first been published in the 10th anniversary year (2018) of the formative framework's creation (2008). It is to be hoped that UK Government appropriately appreciates and values just what it has and continues to do so. Recognising the value and importance of this Act will form the necessary springboard required to propel the nation towards securing, sustaining and improving this pioneering regime wherever possible over the years ahead.

Stirling, UK   Dr. Thomas L. Muinzer
                              Lecturer in Environmental Law and Public Law

ACKNOWLEDGEMENTS

I owe a debt of thanks to numerous people, who have helped me in the preparation of this book in various ways.

Special thanks to Anna-Marie McAlinden, at *Mills Selig* solicitors. At my research institution, Stirling University, I thank in particular Professor Gavin Little, Lorraine Wilson, Dr. Andrea Schapper, Dr. Emily St. Denny and Professor Simon Marsden. Other academics who have offered particularly valuable support include Professor Raphael Heffron, Dr. Darren McCauley, Dr. Peter Doran, Professor Geraint Ellis, Professor Gordon Anthony and Gokce Mete.

At the *European Climate Foundation*, special gratitude is extended to Sharon Turner. I also extend particular thanks to Alina Averchenkova at the *Grantham Research Institute*, Professor Sam Fankhauser at the *London School of Economics* and Jonathan Church at *ClientEarth*. Contact and participation with Stirling University's *Environment and Energy Network*, the Royal Society of Edinburgh-funded *Connecting with a Low-Carbon Scotland Research Network* and King's College London's *Climate Law and Governance* research centre has also been rewarding and beneficial, as has involvement with Matthias Duwe and Lisa Zelljadt at *Ecologic*. I am particulary grateful to Lord Deben, and to those at the Committee on Climate Change that responded to my inquiries.

Many others have been very generous with their time, or have otherwise gone out of their way to provide information, respond to queries, etc., over the course of my work. Particular thanks to Niall Bakewell of *Friends of the Earth (Northern Ireland)*, Sir Crispin Agnew QC, Stephen

Minas, Professor Colin T. Reid, Professor Aileen McHarg, Dr. Ben Christman, Tim Crosland and *Plan B*, Professor David Feldman QC, Professor Richard Macrory CBE, Professor Liz Fisher, Lord Giddens, Professor Benjamin Sovacool and Professor Dieter Helm CBE. At Palgrave Macmillan, I extend particular thanks to Rachel Ballard.

As to others not named above, in particular friends and family, you know who you are and are very much appreciated. The old saying usually attributed to Voltaire springs to mind: 'Appreciation is a wonderful thing. It makes what is excellent in others belong to us as well.'

CONTENTS

CONTENTS

CHAPTER 1

Background to the Climate Change Framework

Abstract This chapter sets out the background to the Climate Change Act, outlining how it has been set in place in order to translate key elements of the UK's national low carbon strategy into legally binding national commitments. An account of the intensive campaigning and political activity that helped to galvanise the Act is provided. The chapter also points out that the framework manifested in law the first long-term legally binding national reduction targets imposed by one nation upon itself in the world and clarifies the reasons why the framework is justly considered to be immensely significant. Environmental governance mechanisms established by the Act are touched on in preparation for Chapter 2, and expert commentary and debate is explored.

Keywords Environmental governance · Climate and energy law and policy · Background to the Climate Change Act
Climate change mitigation and adaptation

© The Author(s) 2019
T. L. Muinzer, *Climate and Energy Governance for the UK Low Carbon Transition*, https://doi.org/10.1007/978-3-319-94670-2_1

1

In the present era, dangers posed by anthropogenic climate change[1] and associated issues (e.g. concerns around energy security[2]) have been driving many states down a path of radical decarbonisation. Against this backdrop, the UK has been undergoing an unprecedented 'Low Carbon Transition'. The UK Low Carbon Transition involves the partial but substantial decarbonisation of the national life, and while it can be understood as an ongoing 'process', it is also articulated in print across certain key government documents, perhaps most importantly UK Government's *UK Low Carbon Transition Plan: National Strategy for Climate and Energy*[3] and *The Carbon Plan*.[4] The UK's Climate Change Act (hereafter 'CCA'), which passed into law in 2008, is an Act of UK Parliament that is designed to articulate in law and achieve in practice the major general processes and outcomes underpinning the Low Carbon Transition. In other words, the CCA is the primary domestic statute[5] that the UK's decarbonisation process is founded on. As the main legislative embodiment of the UK's national climate governance framework, it may come as little or no surprise to scholars and practitioners in the fields of environmental studies, policy, law and governance to find that it manifests a special, complex and frequently challenging legal regime.

[1] See further the body of major reports on this matter issued by the Intergovernmental Panel on Climate Change ('IPCC'), the leading international scientific body convened to examine and assess climate change. In its Fifth Assessment Report, the IPCC has stressed that:

> Human influence on the climate system is clear. This is evident from the increasing greenhouse gas concentrations in the atmosphere, positive radiative forcing, observed warming, and understanding of the climate system. ...It is *extremely likely* that human influence has been the dominant cause of the observed warming [of the atmosphere and the planet] since the mid-20th century.

IPCC, *Climate Change 2013: The Physical Science Basis. Contribution of Working Group I to the Fifth Assessment Report* (Cambridge University Press, 2014), at pp. 15 and 17, respectively (emphasis appears in original).

[2] See C. Mitchell, J. Watson, J. Whiting (eds.) *New Challenges in Energy Security: The UK in a Multipolar World* (Palgrave, Basingstoke, 2013).

[3] DECC, *UK Low Carbon Transition Plan: National Strategy for Climate and Energy* (HM Government, 2009). This was presented to Parliament pursuant to sections 12–14 of the CCA.

[4] DECC, *The Carbon Plan* (HM Government, 2011).

[5] A statute is a written law passed by a legislative body. In the UK, the pertinent legislative body is national Parliament, based at Westminster in London.

The CCA does several large things, and many subsidiary—or smaller—things, the most important of which will be highlighted at several points in this chapter, in advance of the more detailed and in-depth exploration of the framework's mechanisms and requirements to follow in Chapter 2. Most particularly, the CCA imposes rigorous greenhouse gas emissions reduction targets on the UK, doing so in a way where these cumulative targets provide blanket coverage of all of the state's socio-economic sectors. Thus, the framework transposes cornerstone elements of the UK's national decarbonisation programme into legally binding duties: it commits the UK nationally to an 'interim' emissions reduction target, set as a 34% binding greenhouse gas reduction target that is to be met by the date of 2020,[6] and it also commits the UK to a more distant 80% emissions reduction target, set for the date of 2050.[7] These reduction percentages are measured from 1990 baseline emissions levels.[8] The framework also creates and applies a carbon budget system, which introduces finite emission units that are to be steadily reduced over five-year budgeting periods.[9] These target percentages, the carbon budget system and other key mechanisms are outlined and explored in greater detail in Chapter 2. Further, the next chapter will outline and consider the UK's Committee on Climate Change (hereafter 'CCC'), which has also been created under the terms of the Act. The CCC is a public body that is independent of government, which scrutinises and monitors the decarbonisation process and provides expert advice to UK Government and other key governance actors.[10]

As just noted above, the CCA adopts a broad approach that endeavours to provide blanket coverage of all of the state's socio-economic sectors. So, for instance, the CCA has a direct impact upon the transport sector, the agricultural sector, industry, and so on. It will become apparent from the analysis of the substance of the CCA in Chapter 2 that this blanket-style approach can be construed as a 'skeleton' framework that facilitates a potentially broad range of targeted and/or pragmatic political-legal action in the sphere of environmental governance; in realising and employing aspects of these existing capacities for innovative action,

---

[6] CCA, s.5(1)(a).

[7] Ibid., s.1(1).

[8] See further Chapter 2.

[9] CCA, ss.4–10. Discussed in Chapter 2.

[10] The CCC's main advisory elements are set out at CCA, ss.33–35 and s.38.

within the framework's permitted parameters, such action serves to put flesh on the bones of the skeleton framework and contributes towards the achievement of practical outcomes. In addition to blanketing the UK's socio-economic sectors, the framework also blankets the UK *geographically*, that is, the regime covers and applies to the UK as a whole, rather than, for example, being restricted to England and Wales only.[11]

In terms of specific sectoral coverage, this introductory book to the CCA is concerned to hone in most particularly on the sector that has received most attention to date in the context of climate governance, the energy sector.[12] The compound term 'energy sector' can be defined in many ways, and 'energy' itself is a rather elastic and nebulous word that might arguably encompass many things[13]; however, in the present book the 'energy sector' should be understood in a narrow conventional sense, such that it refers to processes, practices and issues pertaining to *the generation of electricity and heat*.[14] Further, 'decarbonisation' of the energy sector refers to the *mitigation or removal of greenhouse gases* from power generation. This said, it will also be seen over the course of this introduction that, while the CCA is heavily concerned with climate change *mitigation*, it is also concerned particularly with the issue of *adaptation*, that is, the need to adapt to the tangible problems and severities arising as a consequence of climate change.

It has been noted in the Preface to this book that the energy sector has been chosen to occupy the central sectoral focus of consideration because energy generation constitutes the single largest source of greenhouse gas emissions in the UK, and it is also the single largest source across the EU and the international community taken generally. Thus, it is the sector that is being most directly and pervasively targeted under the terms of the CCA at the present time, and the decarbonisation of the energy sector stands as a high priority if the Low Carbon Transition is to succeed effectively.

[11] See CCA, s.99, 'Extent'.

[12] See further R. Fouquet (ed.) *Handbook on Energy and Climate Change* (Edward Elgar, Cheltenham, 2013).

[13] 'Energy' can take a wide range of forms, and it is consequently especially difficult to define, see, e.g., J. Andrews, N. Jelley, *Energy Science* (2nd edition, Oxford University Press, 2013), Chapter 1, 'An Introduction to Energy Science', pp. 1–19.

[14] See further, e.g., T.L. Muinzer, G. Ellis, 'Subnational Governance for the Low Carbon Energy Transition: Mapping the UK's "Energy Constitution"' 35(7) *Environment and Planning C* 1176 (2017), p. 1177.

2

It has been highlighted above that the UK has a national Low Carbon Transition strategy in place and that the CCA translates key aspects of this strategy into legally binding national commitments, including the imposition of rigorous greenhouse gas emissions reduction targets upon the UK as a whole.[15] In terms of environmental policy and law, the CCA is an immensely significant legal instrument, for it manifested in law the first long-term legally binding national reduction targets imposed by one nation upon itself in the world. It is not only a complicated piece of legislation in its own right, which contains 101 sections and 8 Schedules and exceeds 30,000 words in length, but the extent of the merits of the legal form in which it has been crafted have been the subject of some disagreement amongst experts, as evoked by the contrast between lawyer Harriet Townsend's suggestion that the Act might be viewed as 'something to be proud of'[16] and lawyer Peter McMaster's view that aspects of the framework are 'legislation at its worst'.[17]

At any rate, and regardless of one's view of technical legal elements or micro-aspects of the overall framework, there is no question that *Friends of the Earth's* description of the CCA as a 'huge step in the fight against climate change'[18] at the time of its creation in 2008 is appropriate. In order to appreciate more fully the framework's values and limitations, and to assess the means by which it came into being and its associated overall sense of development, it is necessary to understand the background to the CCA's creation. It should also be noted that an awareness of the CCA's background and emergence might be drawn on usefully as a means of providing insight and offering lessons to governance actors in other states or within international bodies that are intending to press for the adoption of similar approaches in their respective countries or jurisdictions. Thus, for example, the CCA's emergence might potentially

---

[15] CCA, s.1(1) requires the net UK carbon account for the year 2050 to be at least 80% below 1990 baseline levels; s.5(1)(a) initially set a 26% reduction figure on 1990 levels for 2020, which was subsequently adjusted up to 34% by s.2 of the Climate Change Act 2008 (2020 Target, Credit Limit and Definitions) Order 2009.

[16] H. Townsend, 'The Climate Change Act 2008: Something to Be Proud of After All?' 7(8) *Journal of Planning and Environmental Law* 842 (2009).

[17] P. McMaster, 'Climate Change—Statutory Duty or Pious Hope?' 20(1) *Journal of Environmental Law* 115 (2008).

[18] Public statement released by *Friends of the Earth UK's* climate team in 2008.

extrapolate to some extent to other states that are considering adopting a similar climate governance model; such lessons might prove useful not only for policy makers and government officials, but also for actors within states that might be endeavouring to press governments to engage in robust climate action, such as NGOs[19] and other elements of the general environmental lobby (including sections of the general public).

UK Parliament is 'bicameral', being composed of two houses that jointly create legislation, the House of Commons and the House of Lords. Although the UK is a monarchy, where a Queen or King acts as Head of State,[20] ultimate authority, described in the UK as 'legislative supremacy', is vested in the House of Commons and the House of Lords. Legislation that is 'passed'—that is, created into law—by UK Parliament will typically go through a process where the initial proposed legislation will first exist as a Bill that will be debated and reviewed a number of times, with significant adjustments and revisions commonly being applied over the course of this process, and ultimately the Bill, if it is allowed to proceed, which is not always the case, will become an Act of law. Acts will not necessarily come into force at the moment of their creation, since it is frequently the case that what is known as the 'commencement' of an Act, or parts of an Act, will be delayed, staggered or phased in gradually. In the case of the CCA itself, 'commencement' is dealt with at section 100, and the provisions of the framework came into force over 2008–2009.[21] In general, the nuts and bolts of the Act's regime were up and running relatively quickly.

The CCA was not the UK's first attempt at crafting a climate governance regime, and as such it replaced the UK's pre-existing governance architecture, although that original architecture was considerably less stringent, rather loose, and was founded foremost upon policy. The CCA, by contrast, is a rigorous, binding and coherent legal framework, and has greatly improved upon what had preceded it. Karla Hill has highlighted key features of the UK's pre-CCA climate governance regime in an insightful review of the CCA and its general implications produced for *ClientEarth*, the influential environmental NGO that

---

[19] An NGO is a 'Non-Governmental Organisation', that is, a non-profit group that operates independent of government and that normally seeks to promote some aspect of human (or environmental) welfare.

[20] At the time of writing, the monarch and Head of State is Queen Elizabeth II.

[21] See further the provision made for 'commencement' at s.100 of the CCA.

operates major offices in London and elsewhere internationally.[22] She notes that the UK had adopted a national climate change programme in 2000,[23] which was revised in 2006,[24] and that UK Government had produced energy white papers in 2003[25] and 2007[26] that expressed the need and intention to reduce greenhouse gas emissions from energy, in addition to recognising other problems such as a need for greater energy security and affordability.

Significant recognition concerning the important events that would set the creation of the Bill in motion that was to become in turn the world's most pioneering climate Act must be accorded to the NGO sector. Most importantly, *Friends of the Earth UK* was instrumental in galvanising support for a Climate Change Bill. The primary methods used by the influential environmental campaign group included employing targeted and sustained social campaigning that served to raise consciousness and stimulate support around the issue of the creation of a comprehensive climate framework in the public and non-governmental sector.[27] Simultaneously, high-profile lobbying targeted at UK Government and the political arena more generally ensured that some degree of pressure to take action on the issue was being built up around politicians, government officials and associated powerful interest groups.

Friends of the Earth UK began by pressing for the creation of a comprehensive climate change law in April 2005 and shortly after in May launched a *Big Ask* campaign to push its position. This was augmented by the formation in July 2005 of the *Stop Climate Chaos* coalition, which drew various NGO s together behind a progressive climate mitigation agenda.[28] This *Big Ask* campaign, formed around a coherent,

[22]K. Hill, *The UK Climate Change Act 2008—Lessons for National Climate Laws* (London, ClientEarth, 2009), see 'Background', p. 9.

[23]See DETR, *Climate Change: The UK Programme* (HM Government, 2000).

[24]DEFRA, *Climate Change: The UK Programme 2006* (HM Government, 2006).

[25]DTI, *Our Energy Future—Creating a Low Carbon Economy* (HM Government, 2003).

[26]DTI, *Meeting the Challenge: A White Paper on Energy* (HM Government, 2007).

[27]A summary media and public release document produced by *Friends of the Earth* entitled *The Big Ask Campaign: A Brief History of the Campaign for a New Climate Change Law* is available from the organisation on request and provides a useful chronology of the unfolding campaign.

[28]See N. Carter, 'The Party Politicization of Climate and Energy Policy in Britain', pp. 66–82, in G. Leydier, A. Martin (eds.) *Environmental Issues in Political Discourse in Britain and Ireland* (Cambridge Scholars, Newcastle, 2013), p. 74.

positive and impassioned pro-climate action message, was disseminated and expanded not only through the slight remove of print media, the Internet, and so on, but also through a range of well-organised and participatory public events. The campaign served to create a good focal point that allowed ideas and energies to come together behind a coherent cause, including at the 'grassroots' level through the manner in which it stimulated, cultivated and absorbed growing public interest.[29] *Friends of the Earth* was also able to draw on its relatively well-developed network of members, activists and volunteers across the UK in order to promote and develop the idea across a broad geographic range.[30] These 'bottom-up' aspects of the *Big Ask* campaign were complimented by the sustained lobbying of UK Government by *Friends of the Earth* officials in the more elite spheres of governance, where the group enjoys some significant access and influence in the UK.

Bearing these features in mind, in terms of the campaign elements that helped to galvanise the CCA, it is fair to say that its creation did not come about through the isolated lobbying of political elites, although such lobbying was certainly one distinguishing part of the process; it also drew on 'bottom-up' consciousness raising and mobilisation in a way that served to usefully motivate and gather public interest and action. It is also notable that *Friends of the Earth UK* sustained its campaign pressure over the course of what would become the Climate Change Bill's passage into an Act (discussed below). Again, this included sustained 'grassroots' consciousness raising and bottom-up pressure in addition to sustained high-level pressure focused on UK Government and the elite political arena. In discussing how *Friends of the Earth* pressed members of the public to lobby their MPs[31] in order to advocate for the strengthening of the climate legislation throughout the legislative process, Karla Hill records that by the time the Act was passed approximately 50,000

---

[29] For further discussion of the *Big Ask*, see G. Haq and A. Paul, *Environmentalism Since 1945* (Routledge, Oxford, 2012), pp. 21–22.

[30] *Friends of the Earth Scotland*, which operates more independently than the group's equivalent branches in England, Northern Ireland and Wales, was able to feed the Scottish dimension of the *Big Ask* campaign into the creation of a special subnational Climate Act for Scotland, which enhances the operation of the CCA in that region; this Scottish legislation is examined in detail in Chapter 3.

[31] MPs are Members of Parliament, elected by the public in the UK to represent them in the national Parliament.

people had written to UK Government in support of the legislation.[32] This occurred in a period where the then Labour government was endeavouring to exhibit leadership in engaging with the problem of climate change (including by tabling climate change issues to feature prominently at the G8 summit hosted by the UK at Gleneagles in July 2005), and where the then Conservative Party leader (and subsequently UK Prime Minister) David Cameron was endeavouring to alter perceptions of his own party by augmenting the Conservatives' 'green' credentials.

A further innovative feature of this campaign occurred in early 2008, where *Friends of the Earth* launched the *Big Ask* campaign in a number of countries across Europe in an effort to export UK campaign energy and pressure with positive effect to a more international audience.[33] The organisation projects a good online presence that assists people across Europe in following campaign progress and in getting involved, and asserts that 'Our Big Ask is that governments commit to reduce emissions—year on year, every year'.[34] Given that *Friends of the Earth International* is in effect an international umbrella organisation made up of hundreds of networked environmental groups worldwide, including *Friends of the Earth UK*, its structure and form means that it is well placed to exert a positive influence by these means. The European campaign has helped to encourage progressive climate governance both within and between European states, but it has not gained a degree of traction equivalent to the substantial effect exerted by the *Big Ask* in the UK setting.

Working in conjunction with *Friends of the Earth*, a group of three MPs drawn from what were at that time each of UK Parliament's three biggest parties came together in order to present a model Climate Bill in the House of Commons in April 2005. The MPs were John Gummer (now Lord Deben) of the Conservative Party, Michael Meacher of the Labour Party and Norman Baker of the Liberal Democrats.[35] A parliamentary motion—that is, a proposal that is put forward for debate or decision

---

[32] Hill, supra, n. 22.

[33] See further *The Big Ask   Questions and Answers*, at *Friends of the Earth Europe*, http://www.foeeurope.org/node/670.

[34] Quoting from the 'Big Ask' campaign's homepage, hosted at the *Friends of the Earth Europe*, http://www.foeeurope.org/the-big-ask.

[35] At that time, the Labour Party was in government in the UK, led by Prime Minister Tony Blair.

in UK Parliament—urging the adoption of climate legislation was put forward, ultimately accruing strong support, specifically from 412 MPs out of a total of 646 MPs.[36] The model Bill presented by the three MPs named above at this time was not the actual Bill that directly went on to become the CCA (the UK Government published its own draft Climate Change Bill in March 2007); however, it had a degree of symbolic significance and genuine impact insofar as it introduced a 'tangible' Climate Bill into the debate. Moreover, it did so in a way that, in spite of the controversial nature of the proposals, drew to itself some sense of cross-party unity, given that it had been presented by MPs from UK Parliament's three major parties, who are more often at loggerheads than in agreement on essential issues.

In accordance with the UK's conventional law-making processes, UK Government published a draft Climate Change Bill for public consultation and scrutiny.[37] The consultation was published on 13 March 2007, and it closed on 12 June 2007. In its official response to the public consultation, UK Government noted that '[n]early 17,000 individuals and organisations responded to the public consultation on the draft Bill', adding that '[a]n overwhelming majority of respondents were supportive of the Bill's aim to set and enable the achievement of ambitious emissions reduction targets'.[38] It has been noted above that Karla Hill was commissioned to conduct an independent review of the CCA experience for *ClientEarth* (published 2009). Part of her research included interviews with certain parties that played a significant role over the course of the CCA's development, and aspects of this work help to catch the spirit of the reasons behind the Act's emergence. Based on an interview with an unspecified person or persons from what was then UK Government's Department for Energy and Climate Change,[39] Hill has summarised that:

---

[36] Early Day Motion (EDM) No. 178, 24 April 2005. See also K. Hill, supra, n. 22, p. 9.

[37] HM Government, *Draft Climate Change Bill*, Cm 7040, March 2007.

[38] DEFRA, *Taking Forward the UK Climate Change Bill: The Government Response to Pre-Legislative Scrutiny and Public Consultation* (HM Government, 2007), p. 6.

[39] Hill records that this interview was conducted on 2 October 2009; Hill, supra, n. 22, p. 34, n. 17.

The government considers that the process of pre-legislative parliamentary scrutiny, the public consultation and the [public] campaign were useful and strengthened the [Climate Change] Bill, particularly in relation to the role of parliament and reporting requirements.[40]

She has also summarised as follows from an interview with *Friends of the Earth UK*[41]:

Friends of the Earth considers that the grassroots efforts, based on using its local groups and supporters to build a public campaign as well as lobbying and working with MPs who backed the initiative, were critical in securing the legislation.[42]

Hill adds that '[t]he Big Ask campaign also involved internal learning and capacity building as people working on the campaign developed their knowledge of the issues'.[43] These elements and experiences articulate essential features of the background to the CCA that help to explain the means by which the framework came into being.

A further important stimulus to the Act's creation was provided by *The Stern Report*, which was published in October 2006 and therefore preceded UK Government's publication of the draft Climate Change Bill for public consultation and scrutiny in March 2007.[44] The *Stern Report* was published and launched on 30 October 2006 by a panel that included the report's lead author Lord Nicholas Stern, then Prime Minister Tony Blair and then Chancellor Gordon Brown. It was shortly after this, on 15 November 2006, that UK Government announced that it would produce a Climate Change Bill. While the scientific community had by this time set out the scientific case for redressing climate change in comprehensive terms,[45] and the ethical case was being made strongly by *Friends of the Earth UK*, the *Stop Climate Chaos* coalition and many

---

[40] Hill, ibid., p. 10.

[41] Hill records that this interview was conducted on 7 October 2009; Hill, ibid., p. 34, n. 16.

[42] Hill, ibid., p. 10.

[43] Hill, ibid., p. 10.

[44] N. Stern, et al. *Stern Review: The Economics of Climate Change* (HM Government, 2006).

[45] See amongst numerous scientific examples the body of major scientific reports issued by the IPCC, discussed at supra, n. 1 above.

others, it is well known that most if not all national governments have exhibited a historic tendency to privilege short-term economic considerations disproportionately above other pressing longer-term matters, notably climate change.[46] The *Stern Report* helped to fill a crucial gap in the overwhelming case for action that already existed at that time: it clarified the strong *economic* imperative for engaging in national decarbonisation.[47] As such, in identifying and elaborating a long-term economic case for climate mitigation and adaptation, the *Stern Report* played a vital role in influencing the political class, while also serving to catch the ear in a meaningful way of the corporate and business lobby, and other influential actors with vested economic interests.

Furthermore, and although the corporate and business lobby has tended towards being obstructive of progressive climate governance, with the incidence being particularly acute in the USA,[48] the long-term nature of the carbon budgeting cycles that were due to underpin the framework proposed by UK Government served to some extent to reassure elements of this sector, insofar as they appeared to indicate that a degree of market certainty and stability would be locked-in over time.[49]

[46] Given the USA's role as both the world's most powerful economy and the second largest emitter of greenhouse gasses (behind China), Jennifer Wallace has pointed out how the acute nature of this problem in the USA serves to partially set the tone for the international community; J. Wallace, 'US Environmental Policy and Global Security', pp. 185–195, in N.R.F. Al-Rodhan (ed.) *Policy Briefs on the Transnational Aspects of Security and Stability* (LIT Verlag, Munster, 2007), p. 187. Discussion of this problem in the Australian context arises with reference to climate change in S. Smith, T. Dunne, A. Hadfield (eds.) *Foreign Policy: Theories, Actors, Cases* (3rd edition, Oxford University Press, Oxford, 2016), p. 403. Governmental bias towards short-term economic interests is augmented by the role of powerful corporations that work to downplay the climate problem or outright stimulate support for climate change denial in order to preserve their own short-term economic goals, as highlighted by R.A. Schultz, *Technology Versus Ecology: Human Superiority and the Ongoing Conflict with Nature* (IGI, Hershey, PA, 2013), p. 52.

[47] On this subject in the context of the *Stern Review*, see also N. Stern, 'What Is the Economics of Climate Change?' 2(7) *World Economics* 1 (2006).

[48] On the efforts of ExxonMobil, the Koch brothers and the Donors Trust to use their enormous financial resources to stimulate support for climate denial, often covertly, see Chapter 5, 'Case Studies in the Anthropology of Climate Change', in H. Baer, M. Singer (eds.) *The Anthropology of Climate Change: An Integrated Critical Perspective* (Routledge, Abingdon, 2014).

[49] Carbon budgeting is outlined and discussed in Chapter 2.

Given the West's relatively fundamentalist capitalist approaches to social organisation, Stallworthy has rightly stressed that investors have a key part to play in enabling an adequate response by the state to climate change, pointing out that business and industry likely require environmental policy and law to structure frameworks that not only incentivise investment, but that also maximise investor certainty.[50] The economic uncertainty posed by the potential for change confronting the UK in the lead-up to the CCA's creation was significant; as such, the sense of market certainty and stability represented by the long-term nature of the CCA's carbon budgeting framework had a reassuring effect on elements of the corporate and business sector. Thus, Carter has pointed out that the Corporate Leaders Group on Climate Change seemed to exert some influence on then Prime Minister Tony Blair in seeking for clarity in going forward, even though that clarity itself conceivably could have involved a more challenging emissions reduction framework being locked in place over time: 'Businesses want the stability and certainty to allow them to plan and invest in becoming greener without losing competitiveness'.[51] In a policy report on the CCA incorporating interview findings from various stakeholders, Fankhauser, Averchenkova and Finnegan have recorded that '[t]he energy industry in particular appreciates the legal clarity in the long-term direction of travel' provided by the Act's substantive targets.[52]

The *Stern Report* has also served to highlight the major contribution that environmental economics can make in practice to decarbonisation at the present time; see further the discussion of carbon accounting, emissions trading schemes and associated matters in Chapter 2. Indeed, and bearing in mind that the CCA is a legal framework, in some respects the environmental *economic* dimensions of the climate change challenge are significantly comparable to environmental *law*: particular economic problems can be diagnosed, and economic mechanisms and solutions can

---

[50] M. Stallworthy, 'New Forms of Carbon Accounting: The Significance of a Climate Change Act for Economic Activity in the UK' *International Company & Commercial Law Review* 331 (2007), p. 333.

[51] N. Carter, 'Combating Climate Change in the UK: Challenges and Obstacles' 79(2) *The Political Quarterly* 194 (2008), p. 200.

[52] S. Fankhauser, A. Averchenkova, J. Finnegan, *10 Years of the UK Climate Change Act* (Grantham Research Institute and London School of Economics, 2018), p. 9.

often be devised and applied in practical ways to affect improvements, and likewise, particular legislative (or other legal) inadequacies, gaps or failings can be identified, with opportunities to craft and apply new or developed environmental laws emerging, such that environmental law can contribute to solutions in the area in a similarly practical way.

The introduction of the Climate Change Bill to UK Parliament was announced in the Queen's Speech in 2006 and, as noted above, the government consulted on the draft Bill in March 2007. After the conclusion of this consultation period and the process of legislative scrutiny that runs alongside it, which included the detailed examination and assessment of the Bill by parliamentary committees, UK Government issued its response to this phase in a document entitled *Taking Forward the UK Climate Change Bill: The Government Response to Pre-Legislative Scrutiny and Public Consultation.*[53] After being introduced at Parliament as a Bill in November 2007, the Act received Royal Assent on 26 November 2008, with its major elements entering fully into force by the end of January 2009.[54] The Act as it finally appeared was designed to extend to the UK as a whole,[55] committing the nation to major long-term legally binding greenhouse gas emissions reduction targets, setting a nationally binding reduction target for 2020 at 34% on 1990 levels[56] and an 80% target on 1990 levels for 2050.[57] It has been highlighted above that this stands as the first instance in the world of a nation imposing major long-term binding emissions reduction targets upon itself in law (although other countries have since followed suit; see further Chapter 3).

It has also been highlighted above that the UK is presently undergoing an unprecedented Low Carbon Transition, and the CCA was intended in no small part to cement this process in law. As such, on its creation it became the main legislative embodiment of the UK's national climate governance framework, designed with the intention of pervading emissions outputs at all levels of governance and across the UK's full spectrum of socio-economic sectors. In addition to transposing essential aspects of the national decarbonisation programme into the legally

---

[53] *Taking Forward the UK Climate Change Bill: The Government Response to Pre-Legislative Scrutiny and Public Consultation*, Cm 7225 (The Crown, Norwich, 2007).

[54] See CCA, s.100, 'Commencement'.

[55] Subject to the provisions at CCA, s.99.

[56] Ibid., s.5(1)(a).

[57] Ibid., s.1(1).

binding duties for 2020 and 2050 that have just been highlighted, an inter-related carbon budget system was created at Part 1 to the Act that introduces finite emissions units.[58] The upper emissions range permitted by the budgeting system is to be steadily reduced over five-year budgeting periods. Part 2 to the CCA established the UK's CCC,[59] the non-departmental public body set in place to scrutinise the programme and provide targeted expert advice to a range of key governmental actors, including UK Government and the Devolved Administrations.[60] These features, and other aspects of the substance and detail of the Act, are considered in depth in Chapter 2.

The interim 34% target for 2020 had initially been set at 26%, but it was adjusted upward to 34% in early 2009. This change was applied by a type of law called an Order, specifically the Climate Change Act 2008 (2020 Target, Credit Limit and Definitions) Order 2009, with section 2(2) of the Order making a lasting change to section 5(1)(a) of the CCA, such that the 34% adjustment is locked into the Act in place of the 26% that had previously appeared.[61] The 2050 target has remained at a constant 80% figure. This matter was the subject of some debate at the Bill stage, in particular as to whether the later target ought to sit at 60% rather than 80%.[62] In an interesting development, it was decided to get the CCC that

---

[58] See CCA, Part 1; discussed further in Chapter 2.

[59] That is, the Committee on Climate Change, as introduced earlier above.

[60] Robinson has encouraged the idea that the CCC could be usefully imbued with powers to compel active policy changes; however, this would overreach its fundamental purpose and require substantial adjustment to its overall role, as the CCC has been designed as an advisory and reporting body. See further G. Robinson, 'Stemming the Rising Tide: Developing Approaches for UK Law and Policy to Combat Climate Change' 2(1) *Birmingham Student Law Review* 27 (2017), pp. 28–29.

[61] See further the Climate Change Act 2008 (2020 Target, Credit Limit and Definitions) Order 2009.

[62] The issue of whether international aviation and international shipping emissions should be incorporated directly within the final framework also proved to be a challenging and at times contentious issue. Ultimately, it was decided that they were not to be counted directly within the UK 2020/2050 carbon accounting regime, but a power has been created where the Secretary of State can issue regulations in order to change this circumstance and incorporate them directly if so desired (see CCA, s.30). These emissions sources are, however, still taken into account when budgets are being set and the CCC is providing budgeting advice.

was due to be established under the actual terms of the CCA up and running in advance of the Act's creation, in a modest and less formalised way than would occur when the Act was actually passed. As such, although this committee was formally constituted under the terms of the Act, it existed in a rudimentary form from March 2008 prior to the Act's completed passage through Parliament, and UK Government was therefore able to call upon it to advise on the 2050 target level. In response, the CCC published a formal recommendation letter dated 7 October 2008 stating in no uncertain terms that '[t]he UK should aim to reduce Kyoto greenhouse gas emissions[63] by at least 80% below 1990 levels by 2050'.[64] This recommendation played some role in helping to tip the balance of argument in favour of those advocating for the more stringent reduction figure.[65]

The CCA's development intersected in the UK with the creation of the Department of Energy and Climate Change (DECC), in October 2008. DECC was established in order to address issues and concerns around energy security, energy supply and cost, and climate change within one coherent government department. In effect, it unified climate change policy administration, which had formerly been part of the Department for Environment, Food and Rural Affairs, and energy policy administration, which had formerly been part of the Department for Business, Enterprise and Regulatory Reform.[66] Prime Minister Theresa May closed DECC in July 2016, in a move that generated widespread concern across the environmental sector.[67] UK Government stated that

[63] The reference to 'Kyoto' greenhouse gas emissions alludes to the Kyoto Protocol, an international agreement arrived at in 1997: Kyoto Protocol to the United Nations Framework Convention on Climate Change. The agreement committed the cohort of developed nations listed in Annex I of the United Nations Framework Convention on Climate Change to greenhouse gas reduction targets of a minimum of 5% below 1990 emissions levels over the years 2008–2012. See further the consideration of the international experience in Chapter 3.

[64] Letter from Adair Turner, Chair of the CCC, to the Secretary of State, headed *Interim Advice by the Committee on Climate Change*, 7 October 2008.

[65] The recommendation was based on the consensus view amongst scientists as to what minimum level of action was necessary over the rest of the century in order to most likely mitigate the negative projected impacts of climate change over time.

[66] This became the Department for Business, Innovation and Skills (BIS), which has now developed in turn into the Department for Business, Energy & Industrial Strategy (BEIS).

[67] See further I. Johnson, 'Climate Change Department Closed by Theresa May in "Plain Stupid" and "Deeply Worrying" Move', *Independent* (UK Newspaper), 14 July 2016.

the department's functions would be absorbed into the government's Department of Business, Energy and Industrial Strategy, although this did little to alleviate apprehension in many quarters. The BBC's Environment Analyst, Roger Harrabin, commented that 'DECC has made the UK a world leader in climate policy, and scrapping the department removes the words "climate change" from the title of any department. Out of sight, out of mind, in the basement, perhaps'.[68] However, he also pointed out that: 'here's an opposite scenario: the UK is already bound by its Climate Change Act to step-by-step cuts in greenhouse gases through to 2050'.[69] This captures something of the importance of the steadying influence and long-term sense of certainty provided by the CCA.

3

While the CCA has been rightly praised as a pioneering legislative achievement, it has also been the subject of some debate and critical discussion amongst experts.[70] In order to gain a richer understanding of the framework in general, its particular pros and cons, and the extent to which it might be transplantable to other states or international settings, it is useful to have an awareness and understanding of key aspects of the expert assessment that it has received, and of debate that has arisen in the area. It should be stressed that one is required to tread carefully with the existing critical commentary, which can occasionally be blatantly misleading. A prominent example is Pielke's paper 'The British Climate Change Act: A Critical Evaluation and Proposed Alternative Approach', which according to Google Scholar citation rates (at the time of writing) is amongst the most cited pieces of scholarship on the CCA.[71] Pielke argues that the Act's substantive targets are not realistically achievable,

---

[68]'Analysis—Roger Harrabin, Environment Analyst', set within the news item 'Government Axes Climate Department' by P. Rincon, *BBC News*, 14 July 2016.

[69]Harrabin, ibid.

[70]See, e.g., Macrory's consideration of the CCA in R. B. Macrory, 'The UK Climate Change Act—Towards a Brave New Legal World?' pp. 306–322, in I. Backer, O. Fauchald, C. Voigt (eds.) *Pro Natura* (Universitetsforlaget, Oslo, 2012); Feldman's discussion of the Act in D. Feldman, 'Legislation Which Bears No Law' 37(3) *Statute Law Review* 212 (2016).

[71]R.A. Pielke, 'The British Climate Change Act: A Critical Evaluation and Proposed Alternative Approach' 4(2) *Environmental Research Letters* 1 (2009).

but does not present any meaningful evidence for the proposition, nor does he set out convincingly or in any detail a viable 'alternative approach' (as promised in the title). The paper contains a range of errors, most glaringly where discussion of the important 'interim target' of 34% reductions for 2020 is concerned; the year is consistently stated incorrectly as 2022.[72]

Although the framework has been praised as being pioneering and innovative,[73] certain problem spots have also been identified. In particular, it has been highlighted that the CCA explicitly establishes a range of duties; however, no sanctions are incorporated into the Act for the breach of those duties,[74] and associated concerns around the issues of enforceability and accountability can be seen to arise in a potentially problematic way.[75] Before considering these issues in detail, it will be useful to reflect for a moment on the nature of the duties themselves, and to stress the 'novel' nature of the duties that the Act imposes.

Most particularly, it is the case that the CCA applies legally binding emissions reduction targets that are 'outcome' duties that go beyond the specific legal duties normally imposed on public bodies, thereby raising questions about the nature of those duties themselves. Colin T. Reid has produced insightful commentary that has helped to probe the detail and implications of these matters.[76] Reid has discussed the target duties imposed by the CCA and a similar subnational Act that has been set in place in the UK in Scotland, called the Climate Change (Scotland) Act 2009 (this subsidiary Scottish Act is examined in detail in Chapter 3 of this book). He has highlighted the novel nature of these duties and noted that they can be compared to some extent to the duties that EU environmental law and policy has traditionally applied to EU Member States in

---

[72] Pielke, ibid., pp. 2, 6.

[73] See, e.g., supra, n. 18 (*Friends of the Earth*); M. Lockwood, 'The Political Sustainability of Climate Policy: The Case of the UK Climate Change Act' 23(5) *Global Environmental Change* 1339 (2013), p. 1346.

[74] C.T. Reid, 'A New Sort of Duty? The Significance of "Outcome" Duties in the Climate Change and Child Poverty Acts' 4 *Public Law* 749 (2012).

[75] See further below.

[76] Reid, supra, n. 74. See also C.T. Reid, 'Scotland: Constraints and Opportunities in a Devolved System', in M. Peeters, M. Stallworthy, J.C. de Larragán (eds.) *Climate Law in EU Member States: Towards National Legislation for Climate Protection* (Edward Elgar, Cheltenham, 2012); C.T. Reid, 'Climate Law in the United Kingdom', in E.J. Hollo, K. Kulovesi, M. Mehling (eds.) *Climate Change and the Law* (Springer, Dordrecht, 2012).

order to achieve set results.[77] Over the course of his considerations, he has explained that these sorts of legally binding duties are intended simply to achieve the outcome, so that failure to do so does amount to a breach of the legal duty insofar as Ministers are responsible not merely for doing what they can to achieve it, but for ensuring that it is actually reached.[78] Thus, the CCA, Reid points out, represents an innovation in that the 'duty' is not just to do something, but to also ensure that a specific result or 'outcome' is achieved even though that specified outcome could be affected and dependent on a large number of actors having to do things to help achieve it.

Reid observes that these duties are not wholly unfamiliar, since—as noted above—they share features with obligations that have sometimes been imposed on the UK as a whole under EU law[79]; however, they do make fundamentally new demands on public authorities that are somewhat different from any conventional duties that have preceded them in the context of UK environmental governance (and including beyond environmental governance in most if not all[80] other spheres of UK policy and law).[81] He has concluded that aspects of the broader potential of these novel sorts of duties might be questionable, including particularly questions around the legitimacy of applying long-term duties that endeavour to predetermine policy for future generations; as such, he has speculated that it may be the case that these sorts of duties will endure as a largely one-off response to an exceptional problem (namely climate change).[82] Insightfully, he has also stressed that, regardless of any issues or uncertainty that CCA-style duties might raise, it is the case nonetheless that they create a type of *precedent*.[83] One purpose of this book is to stress that this precedent is not merely a precedent that

---

[77]The EU issues 'Directives' in the sphere of environmental governance (and other areas of governance) that often adopt an approach similar to the CCA. For example, Directive 2009/28/EC, known as the 'Renewables Directive', has applied an overall target to EU Member States requiring an uplift of renewables to 20% of the EU energy share by 2020.

[78]C.T. Reid, supra, n. 74, p. 753.

[79]The UK was an EU Member State at the time of the CCA's creation.

[80]In 2012 Reid pointed out strong similarities existing between aspects of legislative approaches to climate law and child poverty law at that time; see further Reid, supra, p. 74 ('A New Sort of Duty?').

[81]Reid, ibid., p. 749.

[82]Reid, ibid., pp. 766–767.

[83]Reid, ibid., p. 767.

has UK applicability, but one that can be developed and applied any-where in the world if and as desired (of which, see further Chapter 3 on the international experience).

It has been noted above that the CCA explicitly establishes its range of duties, yet at the same time, it does not incorporate sanctions into its provisions for the breach of those duties. In other words, the CCA does not explicitly create or set out any express corrective measures or puni-tive consequences for a breach of the regime's major target (or other) obligations.[84] In the author's view, problems posed by this situation can be usefully explored by dividing the duties imposed by the Act into two different 'types': *substantive* duties and *procedural* duties.[85] The nuts and bolts of the range of actual duties that apply under the framework are addressed in detail in the next chapter, but in terms of principle, *substan-tive* duties should be understood to refer to the active goals and concrete outcomes that are required to be achieved under the terms of the Act, in particular the duties to meet reduction targets, satisfy carbon budget thresholds, and so on. *Procedural* duties should be understood to refer to the automatic processes that are set down by the CCA in order to allow the machinery of the framework to operate, namely obligations concerning reporting duties, consultation duties and similar matters.

Complexity surrounding the extent to which the CCA duties can be enforced, then, evokes a problematic aspect of the framework, and one that has been the subject of significant discussion. The issue has been highlighted by a range of expert commentators, including Stallworthy,[86] Townsend[87] and McMaster.[88] There is no doubt that this issue poten-tially amounts to a serious problem, such that the framework could be arguably much more robust if some form of sanction(s) for breach of the major obligations was included, not least with regard to the

---

[84] As an energy example of sanctions, contrast, for instance, the range of sanctions strung throughout the Energy Act 2011 for failure to comply with aspects of its diverse energy obligations: Energy Act 2011, ss.16, 45, 48, 51, 57, 60, 63, 87B.

[85] In law, substantive duties and procedural duties are often ascribed technical legal meaning. In the present case, the author is using the terms in a generalist, interdisciplinary way, not in a narrow technical legal sense.

[86] M. Stallworthy, 'Legislating Against Climate Change: A UK Perspective on a Sisyphean Challenge' 72(3) *Modern Law Review* 412 (2009).

[87] Townsend, supra, n. 16, p. 842.

[88] McMaster, supra, n. 17.

milestone emissions reduction targets for 2020 and 2050. States seeking to adopt a CCA-style climate regime might wish to learn from this arguable shortcoming and construct clear sanctions that penalise a failure to meet the framework's duties. Nonetheless, the extent of this problem should be kept in perspective. In the first place, it should be emphasised that the framework is *legally binding*, and as such, the law *does* insist that its requirements are to be met. This circumstance can be contrasted with a 'policy-oriented' climate framework, which might sketch out policies and intentions, but which will lack the binding weight of law behind it in order to support its goals.[89] In other words, simply because environmental law does not sketch out particular sanctions that the courts might apply in the case of a breach of a particular duty, this does not mean that the duties themselves are not legally binding; governmental actors *are* obligated to endeavour to abide by them. As such, a framework exhibiting these sorts of characteristics can still function well. For instance, and to take an energy-specific example, it is the case that under the Energy Act 2016 the UK's Oil and Gas Authority has a range of important specified functions.[90] If certain functions are not adequately fulfilled, there is a type of 'soft sanction' that can be taken against the Authority; however, this is basically a process of review and then guidance and a test for fitness of purpose[91]; it is not a meaningful conventional legal sanction. This does not mean, however, that the framework ceases to bind on pertinent actors, or that it may be disregarded or casually ignored: it retains legal force and has operated tolerably well to date in practice.

Nor do these circumstances mean necessarily that there are no avenues that might be pursued in order to access some degree of justice if a substantive (or indeed procedural) duty is blatantly disregarded. Reid has emphasised that judicial review provides an available route in the UK for seeking a sanction or a remedy in the courts.[92] Judicial review amounts to a process of constitutional protection in the UK, where the courts are imbued with the authority to safeguard the public interest by protecting against aspects of the abuse of power by public authorities.[93]

---

[89] In effect, this was the case in the UK prior to the CCA's creation.

[90] See Energy Act 2016, ss.1–7.

[91] Energy Act 2016, s.16.

[92] Reid, supra, n. 74, p. 757.

[93] For more detail, see 'What Is Judicial Review, and What Are Its Purposes?' pp. 2–6, in G. Anthony (ed.) *Judicial Review in Northern Ireland* (Hart, Oxford, 2014).

In other words, where the rights of citizens are impacted negatively by the inappropriate exercise of public authority under the terms of the CCA, judicial review gives UK courts some ability to ensure that the duties imposed by Parliament are appropriately interpreted and abided by.[94] Therefore, in circumstances where there has been a failure to achieve CCA substantive duties or to abide by procedural duties, a potential avenue for redress is provided by application for judicial review. However, Reid has also pointed out that it is unlikely that any individual citizen could claim compensation through this channel, due to what he describes as *the unlimited class to whom the duty is owed*, that is, the UK public in general rather than targeted individual citizens or distinct sections of the public.[95] This barrier to compensation, he adds, is further compounded by the problem of how *specific* sorts of losses resulting from a failure to achieve a substantive duty like a major CCA reduction target could be attributed to an individual or group of individuals in any sort of narrow or concrete way.[96] There is no doubt that this is the correct position.

It should be noted that although the form and substance of the CCA is novel and pioneering, this type of issue in itself is in some respects nothing new to UK environmental governance, insofar as the breaking of substantive target duties does not normally give rise to remedies in public or private law.[97] While it is debatable as to whether or not this situation could be improved upon, on reflection it is far from wholly unreasonable: CCA-style substantive target duties by their very nature are not designed to give individuals strong causes for action in courts of law or elsewhere, and damages are not normally intended to be available for individuals under this type of schematic. It would be an odd situation, for instance, if some specific individual, group of individuals or categorisable section of the public achieved a windfall payment from the public purse as a consequence of UK Government failing to satisfy a substantive CCA carbon budget. Other private individuals might feel aggrieved that they did not receive some form of compensation as well, or indeed

---

[94] See further M. Fordham, *Judicial Review Handbook* (6th edition, Hart, Oxford, 2012).

[95] Reid, supra, n. 74, p. 757.

[96] Reid, ibid.

[97] See further C. Callaghan, 'What Is a "Target Duty"?' 5(3) *Judicial Review* 186 (2000); *R v London Borough of Islington ex p. Rixon* [1997] ELR 66, 69.

that their own public taxes could be forming the basis of the payout. In terms of this latter point, it is to be remembered that one is dealing with national government itself under the CCA's approach (not, e.g., private corporations).[98] In sum, courts will tend to hold back from enforcing substantive 'target duties' of this sort, which by their very design have a somewhat aspirational character.

This said, these sorts of complications surrounding enforceability and compensation do not necessarily mean that a private individual or individuals would be prohibited from endeavouring to enforce the CCA's duties in court, and with some resulting degree of success awaiting them at the end of the process, given that the duties are expressed under the terms of the Act in a clear and binding manner. This is most evidently the case where *procedural* duties are at issue: the general position is that legal challenges for a breach of duty may be potentially successful where they are intended to secure compliance with the procedural sorts of duties noted above (reporting and consultation duties, and similar).[99] Where *substantive* duties are at issue, however, namely reduction targets, carbon budget thresholds and associated matters, a concrete degree of success is considerably less likely. In these latter sorts of cases—for example, where the 2050 reduction target has been significantly missed—it is likely that the courts will award 'declaratory relief', meaning in essence that they will declare that there has been a breach of the legal obligation in question under the terms of the CCA. This may likely be an outer limit of what the court will do. In addition to the existing work on these sorts of matters from lawyer Colin T. Reid, further commentary from legal expert Aileen McHarg confirms this position.[100] In effect,

[98] See further the consideration of the role of UK Government and the Secretary of State under the terms of the framework in Chapter 2.

[99] This general assumption in itself is also subject to uncertainties, however. Feldman has suggested that where the Secretary of State fails to produce an annual report to Parliament in accordance with mandatory CCA reporting duties, the courts may be unlikely to issue a mandatory order for the report to be produced due to the fact that they do not normally make orders that intrude on the working of the Houses of Parliament; for this proposition, Feldman cites the Bill of Rights 1688, Art. IX as legal authority. See further Feldman, supra, n. 70, p. 222.

[100] A. McHarg, 'Climate Change Constitutionalism? Lessons from the United Kingdom' 2(4) *Climate Law* 469 (2011), p. 477.

this sort of declaratory relief is merely a statement of the obvious, where the court declares that non-compliance has occurred. Beyond this, it is unlikely that remedies of a more robust nature will be forthcoming.[101]

Other states or policy makers seeking to draw on the UK framework, or aspects of it, are well advised to be aware of these issues: Reid has noted that there seems to be a sense of 'paradox' here, where sharply drawn legally binding duties are imposed on UK Government in a way that serves at the same time to restrict the way in which those duties can be enforced.[102] It might be concluded that this amounts to an acceptable design complication that is necessary to frameworks of this nature; or, conversely, it might be concluded that these matters embody unacceptable problems that require the development of new or novel solutions. As Stallworthy has suggested, the courts' role will likely be limited, and one cannot realistically expect them to apply substantial constraints to the major issues.[103] Alternatively, it might also be the case that where a CCA-style approach is being adopted by a different state, the state in question might already have some form of solution that exists within its pre-established environmental law and governance regime. It is certainly the case that where other states seek to adopt a CCA-derived approach to climate governance, the framework will need to be adapted to fit the concrete national circumstances in which it is being applied; as such, it may be possible to derive or generate particular solutions to the sorts of problems arising from the issues that have just been sketched out from the pre-existing governance context in question. Thus, in 'The Legally Disruptive Nature of Climate Change',[104] Fisher, Scotford and Barritt

---

[101] Though note that work from the author has argued that there *is* some facility under the terms of environmental law for courts to take a more robust approach to enforceability in the express context of the CCA than the law may have traditionally permitted, given both the innovative nature of the CCA itself and the novel nature of the issue it is designed to combat (climate change): T.L. Muinzer, 'Is the Climate Change Act 2008 a "Constitutional Statute"?' *European Public Law* (2018, forthcoming). See also J. Church, 'Enforcing the Climate Change Act' 4(1) *UCL Journal of Law and Jurisprudence* 109 (2015).

[102] Reid, supra, n. 74, p. 762. A related but slightly different position might be that these sorts of environmental target duties might be regarded by the courts as a mandatory relevant consideration, so that a failure by the government to take them into account when making policy would be an error that the courts could review. The author is grateful to Professor David Feldman QC for raising this point over the course of research.

[103] Stallworthy, supra, n. 50, p. 340.

[104] E. Fisher, E. Scotford, E. Barritt, 'The Legally Disruptive Nature of Climate Change' 2(80) *Modern Law Review* 173 (2017).

have highlighted how the Dutch courts have successfully developed the doctrine of 'hazardous state negligence'[105] and the New South Wales Land and Environment Court in New Zealand has drawn upon the principle of 'intergenerational equity'[106] in order to deal with novel problems arising through climate change challenges.

It is also necessary to keep the significance of issues around enforceability and sanctions (and compensation) in perspective. The most essential issue predominating in this area can be summarised as follows: the CCA establishes long-term legally binding emissions reduction targets and underpins them with a binding carbon budgeting system, but it does not include sanctions that can be applied where UK Government fails to ensure that its substantive targets are met.[107] Consequently, the Act has drawn criticism from experts, notably from McMaster in 2008, who described it as 'legislation at its worst' when it was at the Bill stage due to these sanction omissions.[108] While attention to these legitimate concerns is warranted, and while McMaster's point was valid insofar as he was stressing that the government should have included explicit sanctions in the legislation if, as it claimed, it intended some degree of sanction to be available,[109] his characterisation taken in isolation constitutes extreme hyperbole ('legislation at its worst'), given that the CCA was the most forward thinking and important national climate statute ever created at the point of its genesis. It has been emphasised earlier in this chapter that one of the CCA's major contributions to environmental governance is the manner in which it manifested in law the first long-term legally binding national reduction targets imposed by one nation upon itself in legislation in the world. In addition to articulating and critiquing the CCA's shortcomings, which fulfils a valuable function, it is important that analysis within the field of environmental studies also recognises its merits proportionally, such that strengths can be identified and built upon clearly and lucidly.

[105] See *Urgenda Foundation* v *Netherlands* (24 June 2015) ECLI:NL:RBDHA:2015:7196.

[106] *Landscape Guardians Inc* v *Minister for Planning* [2007] NSWLEC 59; see further the discussion at Fisher, Scotford and Barritt, supra, 104, p. 191.

[107] See further H. Townsend, supra, n. 16, p. 842; and (unattributed) 'Energy Legislation: The Climate Change Act 2008' (legislative comment) *Environmental Law Monthly* 1 (2008).

[108] McMaster, supra, n. 17.

[109] UK Government had indicated that judicial review would be available to operate as a form of sanction, although it was not expressly stated in the legislation.

While concerns over an inability to impose sanctions are countered to an extent by the framework's in-built facility for judicial review, imperfect as that medium may be, such concerns are also countered to some extent by other subtle features. These include targeted procedural duties, in particular the manner in which the framework builds in a rigorous reporting system,[110] where, amongst other obligations, the Secretary of State is required to lay major progress reports before UK Parliament on a regular basis.[111] Constant scrutiny and the high visibility of progress levels thereby place a general pressure upon national government to ensure that the framework's substantive (and indeed procedural) commitments are met.[112] This pressure is reinforced through a variety of means, including negative coverage of governmental failures in the media, criticism levelled at the government in the policy arena by opposing political parties, and the pressure actions of NGOs like *Friends of the Earth*, *ClientEarth* and others.

In addition to the concerns just raised, a further and partially related concern that has also crystallised in expert commentary pertains to the issue of *accountability*. Reduced to its essentials, problems in this area concern the extent to which blame and responsibility for failures to meet substantive duties under the CCA can be apportioned directly and actively to appropriate actors. This in turn raises issues about the culpability of public authorities, and indeed governmental liability more generally insofar as UK Government has primary responsibility under the terms of the CCA for achieving the framework's major objectives.[113]

---

[110]The CCA's reporting obligations are outlined in Chapter 2.

[111]See Chapter 2.

[112]Townsend has also made the following salient observation:

> Any law student could tell Government that a duty of this kind is simply unenforceable but, as Ministers insisted during the Bill's passage through Parliament, they might just be missing the point. The Government's objective in setting this statutory target is to give greater certainty as to the scale of change required than could be achieved by more traditional means. In doing so they seek international credibility for the UK in treaty negotiations, and domestically the platform for a gear change in policy formation.

H. Townsend, 'Climate Change Act 2008: Will It Do the Trick?' 11(2) *Environmental Law Review* 116 (2009), p. 117.

[113]See further Chapter 2, and in particular discussion of the role of the Secretary of State in UK Government.

Callaghan has highlighted that courts in the UK will be unlikely to find a public authority to be in breach of a target duty automatically even where that duty has not been fulfilled, but will consider if the public authority is doing its best to put the failure right; if it is doing all it reasonably can, the courts will be unlikely to intervene.[114] Courts will also take general practicalities into account, including budgetary constraints, scarcity of necessary resources, the lack of a governmental actor's room to manoeuver due to the permitted political behaviour of other governmental institutions, and so on.[115]

Given that the Secretary of State from UK Government has primary responsibility under the terms of the CCA for ensuring that the major emissions reduction targets and associated major substantive duties are met,[116] where a failure to meet those duties might be strongly influenced, shaped or determined by broader governmental actors— for example, by budgetary constraints imposed by the Treasury— the courts will make some allowance for these sorts of constraints in law. These observations suggest that the UK's pioneering framework *appears* to imbue the Secretary of State with an ultimate onus and facility to act in this area, most particularly through the way in which he is positioned as the governance actor with primary authority and responsibility; however, this might amount to less than may be expected if influential government departments and associated key actors are not acting in relative harmony (or, indeed, if a coalition government arises, as occurred in the UK in 2010 when the Conservatives and Liberal Democrats formed a coalition government). A general lesson can be seen to lie behind the framework: it will be at its functional best where governance actors endeavour to act in concert towards its substantive ends, and where there is division and tension, this will tend towards acting as an undermining force.

---

[114]Callaghan, supra, n. 97, pp. 185–186.

[115]On the courts making allowance for the Treasury's budgetary control in assessing the Secretary of State's role in constricting solar energy subsidies, see T.L. Muinzer, '"To PV or Not to PV": An Analysis of the High Court's Recent Treatment of Solar Energy' 17(2) *Environmental Law Review* 128 (2015).

[116]'It is the *duty of the Secretary of State* to ensure that the net UK carbon account for the year 2050 is at least 80% lower than the 1990 baseline' (CCA, s.1(1), emphasis added), etc.

4

Another dimension of the CCA that has formed a basis of serious discussion and debate amongst experts relates to environmental constitutionality/constitutionalism. This sphere of critique has tended to centre on the relationship between the CCA and its constitutional environment, that is to say, the way in which the framework has been (and continues to be) influenced or otherwise impacted by the UK's constitutional setting. In this area of thinking, 'constitution' in effect refers to the way in which the governance of the state is structured,[117] or what Turpin and Tomkins describe as the body of rules that govern the political system, the exercise of public authority and the relations between citizen and the state.[118] Indeed, it is notable that some of the considerations above concerning the nature of the CCA's duties, and associated issues including enforceability and accountability, have cohered in scholarly commentary around these sorts of additional concerns relating to 'constitutionality'.

Thus, the CCA's relationship to constitutional matters has received enough attention, and is of enough importance, to render the framework's relationship to the national constitution as an important and reasonably well-developed concern. Further, while Chapter 3 in this book deals with the broader international implications of the CCA, it is worth stressing at this point that the framework's relationship to its constitutional environment is not merely a matter of UK-specific importance; where other states may seek to adapt the framework or allow it to exert practical influence on governance design in other state settings, it should be borne in mind that the regime's relationship to its constitutional context will no doubt have significant implications for its operation. This will be seen more clearly through concrete reflection on the UK experience.

First, it is to be noted that the UK does not have a 'written constitution', that is, a conventional constitutional document that draws together the state's core constitutional principles.[119] Here, the UK

---

[117]See generally M. Loughlin, *The British Constitution: A Very Short Introduction* (Oxford University Press, Oxford, 2013).

[118]See C. Turpin, A. Tomkins, *British Government and the Constitution* (6th edition, Cambridge University Press, Cambridge, 2007), p. 3.

[119]For more on this, see N. Parpworth, *Constitutional and Administrative Law* (8th edition, Oxford University Press, Oxford, 2014), p. 11.

can be contrasted with its neighbours the Republic of Ireland[120] and France,[121] and the numerous other countries around the world that have constitutions that are codified in this manner. Instead, in the UK's case, a variety of disparate sources containing parts or aspects of the UK constitution can be identified, a sort of patchwork that can be linked together to add up to a fulsome constitution. The range of sources serving to manifest elements of the UK's constitution includes relatively concrete things like Acts of UK Parliament and rulings of the courts that display a constitutional character. The sources also include 'principles', which are somewhat more abstract in nature, most particularly the principles of *parliamentary sovereignty* and *the rule of law*, where the first principle recognises an ultimate level of authority vested in the democratically elected national Parliament, and the second principle in essence recognises that everyone is equal before the law. Further UK constitutional sources include features like the royal prerogative powers, which are a limited range of powers that are normally held symbolically by the Monarch and exercised in practice by UK Government under the government's executive capacity for action (i.e. without requiring the direct consent of Parliament), and constitutional conventions, which are rules of government that are observed chiefly as a matter of custom.[122]

Fundamentally, a state's constitution normally serves to structure the major governance institutions of the state, articulate the balance of power running across that governance machinery, and map out the parameters of the rights and duties enjoyed by individual citizens in relation to it. In terms of environmental law and policy, Elliott and Thomas have summarised that 'constitutional law' is most particularly concerned with the sorts of basic ground rules that determine both the government's powers and the fundamental rights enjoyed by individuals[123]; these matters clearly interact profoundly with environmental governance in the context of state power, environmental liability and environmental accountability.

---

[120] Constitution of Ireland—*Bunreacht na hEireann.*

[121] Constitution of France. The present version dates from 1958 and has been amended on a number of occasions.

[122] See further Parpworth, supra, n. 119, p. 12 ('Sources of the UK Constitution').

[123] M. Elliott, R. Thomas, *Public Law* (2nd edition, Oxford University Press, Oxford, 2014), p. 4.

In an important critique of the CCA entitled 'Climate Change Constitutionalism?' (2011),[124] Aileen McHarg has enquired into the extent to which the CCA framework itself may or may not be understood to have some degree of 'constitutional' status. Investigation into the extent to which the CCA might be constitutional in nature engages with important issues, because, as McHarg has pointed out, broad environmental values are often found in constitutional documents,[125] but in the case of the UK, there is no 'codified' constitutional document to turn to. In the present writer's view, it is also significant that where one endeavours to find a concrete list of explicit constitutional *rights* enjoyed by private citizens in the UK, one commonly turns to an important constitutional statute, the Human Rights Act,[126] which sets out the core range of explicitly protected rights.[127] An explicit 'environmental' right, however, receives no mention whatsoever; for example, something that might be phrased as a 'Right to a Clean and Safe Environment', or similar.[128] As such, an explicitly articulated 'environmental' right seems to be lacking in the UK constitutional context. Therefore, in investigating whether the CCA is in some sense an explicit part of the UK 'constitution', or whether it can be said to attract to itself some sort of constitutional 'status', one is simultaneously raising a question of much broader importance, namely: do the important environmental protections that a CCA-style approach endeavours to afford embody something that amounts to a level of UK constitutional environmental protection in the area of climate change?

McHarg persuasively links climate change protections as intersecting with the sorts of environmental values one often finds in constitutional documents.[129] Over the course of discussions engaging with 'environmental constitutionalism', she suggests that it is arguable that the approach to the prevention of dangerous climate change embodied by the CCA might realistically be viewed as one form of constitutional

---

[124] McHarg, supra, n. 100.

[125] McHarg, supra, n. 100, p. 476.

[126] Human Rights Act 1998.

[127] Ibid. (Human Rights Act); see particularly the rights set out at Schedule 1 to the Act, which are drawn from the European Convention on Human Rights.

[128] See ibid. (Human Rights Act 1998).

[129] McHarg, supra, n. 100, p. 476.

expression within environmental constitutionalism's broader sphere.[130] She clarifies that the uncodified nature of the UK constitution arguably creates conditions where aspects of environmental constitutionalism might arguably be seen to manifest in the UK's constitutional setting through the CCA. As such, the framework might be viewed conceivably as an expression of environmental constitutionalism (to at least some extent), albeit an incomplete one.[131] McHarg's research indicates that it is possible to credibly view the CCA from some sort of general constitutional perspective. As such, McHarg appears to lean towards the view that the CCA is at least in some respects 'constitutional', with her critique demonstrating how a good case might be made for this proposition. She suggests that 'as a pre-commitment strategy designed to promote the long-term public interest in GHG[132]-emission reduction by constraining short-term political and economic imperatives, the CCA can reasonably be described as a constitutional measure'.[133]

In a study entitled 'Is the Climate Change Act 2008 a "Constitutional Statute"?', the author has elaborated how the CCA within the UK, and by logical extension any similar or analogous major derivation or extrapolation of that framework internationally, should be broadly interpreted or construed as being 'constitutional' in nature.[134] This is due to the fact that it is a primary legislative framework designed to counteract anthropogenic climate change, an environmental problem so potentially severe and of such pervasive present and future public importance that the correct rational position is that its redress must feature highly on any list of core rights, protections and entitlements. Noam Chomsky has emphasised that the vast majority of scientists, the IPCC,[135] and all of the world's greatest national academies of sciences and professional societies of science are agreed that global warming is taking place, that a substantial human component is contributing to it, that the situation is dangerous and that the world is potentially moving towards a tipping point

---

[130] McHarg, ibid., pp. 476–477.

[131] McHarg, ibid., p. 476.

[132] Greenhouse Gas.

[133] McHarg, supra, n. 100, p. 483.

[134] T.L. Muinzer, supra, n. 101 ('Is the Climate Change Act 2008 a "Constitutional Statute"?').

[135] The IPCC.

that could escalate the process in a swift and possibly irreversible way.[136] Bearing in mind that the CCA is intended to make a significant contribution towards redressing these problems, the correct rational position is that a default assumption must exist such that the framework should be interpreted as having some significant degree of elevated constitutional status.

While it is submitted that these points reflect the appropriate conclusions to be drawn from a general rationalist consideration of the important ethical issues attaching to the CCA in the context of constitutionality and environmental studies, the author's study[137] has also addressed certain technical legal concerns arising within the field of environmental law and constitutional law. The critique endeavours to provide a technical legal analysis that demonstrates that it is possible to frame the CCA's sense of constitutionality in narrower legalistic terms. The study points out that since an influential legal judgment delivered in the case *Thoburn v Sunderland City Council* (2002),[138] it has become increasingly common in UK legal analysis to construe Acts of UK Parliament as being either 'ordinary statutes' or 'constitutional statutes'. The study also points out that 'constitutional statutes' are understood to attract a special legal status in law.[139] This begs the question of whether the CCA itself can be classified as a 'constitutional statute', and as just noted, the study provides a technical legal analysis contending that it is possible to answer this question in the affirmative.[140] Where the framework *is* understood in this way, it is possible to proceed to calculate and sketch out specific legal outcomes or consequences that might arise in law where the Act is treated as a 'constitutional' statute by the UK courts.

---

[136]Noam Chomsky made this statement in a public lecture, entitled 'Global Warming and the Common Good' (delivered at East Stroudsburg University, February 7, 2013). The talk is transcribed on the *Reading Chomsky* website, where this statement is drawn from, see, http:// readingchomsky.blogspot.co.uk/2013/04/normal-0-0-2-false-false-false-en-us-ja.html.

[137]Muinzer, supra, n. 101.

[138]*Thoburn v Sunderland City Council* [2003] QB 151.

[139]See, e.g., Lord Hope's treatment of the Scotland Act 1998 in *H. v Lord Advocate* [2012] UKSC 308, at para [30].

[140]While it is possible to conclude that the CCA is a 'constitutional statute' in these terms, the courts may not necessarily take this view. A case has not yet arisen to resolve the point.

The general position, as set out in that study, is that where the CCA is accepted as a constitutional statute, it will be likely subject to a greater degree of entrenchment than an 'ordinary' statute, meaning that it will be likely harder to repeal or replace it in comparison to regular Acts.[141] It also appears to be the case that this sort of constitutional statute designation will potentially impact the courts' interpretive treatment of the CCA's substantive duties and the remedies that might be applied by the courts in the event of a breach of those duties. Technical legal reasons are offered as to why the courts might adopt more robust remedial approaches where it is clear they are dealing with the breach of a 'constitutional statute' rather than an 'ordinary statute'.[142] In general terms, it has been demonstrated that the interpretation of the CCA as a constitutional statute has a credible basis in UK law, and it has been pointed out that this sort of interpretation sets the Act on a robust footing appropriate to its novel design, content and objectives, while also potentially serving to reinforce or even expand the courts' scope in dealing with breaches of the Act's duties somewhat. This is clearly significant, for, as emphasised at multiple points across this chapter, one particular area of contention concerns the fact that the CCA creates duties but neglects to incorporate sanctions into its framework for the breach of those duties.[143] The question of whether or not the CCA actually *is* a 'constitutional statute' in law is a question that can be answered in practical terms by the courts, but a legal case raising the issue has not yet come before the courts in order to compel the answer at the time of writing.

5

A further significant concern raised by critique of the CCA relates to the national framework's relationship to 'higher' levels of governance, e.g. the international level,[144] and 'lower' levels of governance, often generalised for rhetorical convenience as the 'subnational' level within the UK. On its creation into law, the CCA and the policy trajectory that it

---

[141] See generally, T.L. Muinzer, supra, n. 101.

[142] Muinzer, ibid.

[143] Muinzer, ibid.

[144] Another example of a 'higher' level would be the EU/'supranational' level, during the period of the UK's membership of the EU.

manifested were generally perceived as avant-garde and forward thinking. Shortly after the passage of the Act, legal expert Harriet Townsend characterised the legislation as embodying arguably 'an old-fashioned, some would say British, aspiration to do the right thing' that the UK could 'be proud of'.[145] Indeed, and bearing in mind this sort of cumulative 'British' evocation of a national framework that spans the UK as a whole, it is also worth highlighting that a conventional 'national' interpretation of the framework tells only one part of the story. In the first place, while the CCA is indeed a 'national' innovation first and foremost, it has been highlighted that the Act initially functioned as the partial implementation answer to supranational legislative and policy drivers, due to the UK being a Member State of the European Union at the time of the Act's creation.[146] The CCA did not directly transpose or apply EU law within the UK, and as such, it was a free-standing national innovation; however, it was designed with a degree of flexibility in order to accommodate and absorb requirements under the EU's Emissions Trading Scheme and associated obligations.[147] This EU experience was in turn interconnected to the international climate governance experience.[148]

Furthermore, by the time 5 years had elapsed since the Act's creation, scholarly research was drawing attention to the fact that the common, intuitively credible way in which the CCA's form, objectives and practices tend to be viewed in 'national' terms is in no way fully reflective of the underlying constitutional reality of UK climate law: one cannot strictly speaking conceive of the UK framework as a 'State' endeavour without accounting to some extent for the presence and influence of UK devolution. In highlighting these elements in their work, Sharon Turner

---

[145] Townsend, supra, n. 16, p. 842.

[146] See further the detailed analysis of EU-UK governance dynamics in this area in T.L. Muinzer, 'An Evaluation of the Implications of EU Climate and Energy Governance for the UK in Light of Brexit' 23(2) *European Journal of Current Legal Issues* (2017).

[147] These sorts of multilevel dimensions are outlined and examined in T.L. Muinzer, 'Does the Climate Change Act 2008 Adequately Account for the UK's Devolved Jurisdictions?' 25(3) *European Energy and Environmental Law Review* 87 (2016).

[148] Muinzer, supra, n. 146. Benjamin has suggested that due to the CCA's sense of alignment with international and EU obligations and developments, it can be viewed as incorporating a 'transnational' dimension; L. Benjamin, 'The Responsibilities of Carbon Major Companies: Are They (and Is the Law) Doing Enough?' 2(5) *Transnational Environmental Law* 353 (2015), p. 360.

(who honed in most particularly on Northern Ireland)[149] and Colin T. Reid (who focused most particularly on Scotland)[150] in effect pointed out that there is a rich, powerful and complex governance experience playing out *beneath* the national level within the UK's devolved jurisdictions, which carries the implication in turn that this subnational arena may be liable to significantly impact the 'national' framework's processes and intended outcomes. Indeed, given the perceived uncertainty surrounding the role and influence of devolution at that stage of the CCA's development, a sense emerged that the UK's devolved jurisdictions would potentially play a major role in determining whether the UK could actually achieve the substantive targets it had set for itself, in spite of the targets' 'national' nature.[151] These points are especially important insofar as the extent to which EU and international developments have had an impact on the UK climate regime tends to be overtly recognised and addressed, whereas the subnational dimensions of the CCA have tended more often than not to be somewhat hidden or eclipsed by the 'national' features of the framework and the associated perceptions attaching to it.

The author has since been funded through Queen's University Belfast (QUB) by the Northern Irish government's then Department for Employment and Learning[152] to investigate the impact of devolution on the CCA, with the research taking place over 1 October 2011–1 July 2015 at QUB's School of Law.[153] Also since that time, the UK's Economic and Social Research Council has funded a cohort of experts to examine the relationship between the UK renewable energy transition

---

[149] S. Turner, 'Northern Ireland's Consent to the Climate Change Act 2008: Symbol or Illusion?' 25(1) *Journal of Environmental Law* 63 (2013); S. Turner, 'Committing to Effective Climate Governance in Northern Ireland: A Defining Test of Devolution' 25(2) *Journal of Environmental Law* 203 (2013).

[150] Reid, 'Scotland: Constraints and Opportunities in a Devolved System' (supra, n. 76), p. 137; Reid, 'A New Sort of Duty? The Significance of "Outcome" Duties in the Climate Change and Child Poverty Acts' (supra, n. 74); C.T. Reid, 'Climate Change Law in Scotland' Ympäristö-Juridiikka Miljöjuridik (1) *Finnish Environmental Law Review* 18 (2012). See further Reid, supra, n. 76 ('Climate Law in the United Kingdom').

[151] See Turner, supra, n. 149, 'Committing to Effective Climate Governance in Northern Ireland'; Reid, supra, n. 76 ('Scotland: Constraints and Opportunities in a Devolved System').

[152] The department's functions have since been transferred to the Northern Ireland Executive's Department for the Economy and Department for Communities.

[153] T.L. Muinzer, *The UK's Energy Decarbonisation Process and the Challenges of Devolution* (PhD research thesis), Queen's University Belfast (1 October 2011–1 July 2015).

and devolution. This 'Delivering Renewable Energy Under Devolution' group (and its associated research outputs) is commonly identified with the project's acronym, DREUD.[154] The group exercised a particular (but by no means exclusive) focus on Scotland, and in doing so the DREUD research has shed useful light upon the relationship between substate-level Scottish and national-level UK energy governance, while also casting valuable light on the subnational governance experience more broadly.[155]

In general, these research investigations have found that the realistic and logical view is that the subnational sphere is highly constrained by the national sphere, but that the subnational sphere does have some significant capacity for autonomous action nevertheless, and, as such, it therefore follows that the subnational sphere, including powerful subnational governance actors, has a key role to play in working towards the attainment of the CCA's 'national' objectives, if those objectives are to be realised as fully as they are intended to be under the terms of the CCA. On inquiry, the author has determined that one cannot come remotely close to assessing in any broad sense the extent to which key governmental actors in the devolved sphere of governance[156] are subject to opportunities and constraints on their overall capacities for action by examining in isolation the room to manoeuvre accorded either explicitly or implicitly under the terms of the CCA itself. Rather, in conjunction with exploring those dimensions of the CCA framework, one must also (at the least) investigate and assess the extent to which capacities and constraints act upon subnational governmental action under the terms of the UK's devolution arrangements.[157] A recognition of these necessities means that hard problems open up for environmental studies

[154] 'Delivering Renewable Energy Under Devolution', funded by the ESRC, commencing in January 2011 and finishing at the end of January 2013.

[155] Important DREUD research outputs include: G. Ellis, R. Cowell, F. Sherry-Brennan, P. Strachan, D. Toke, *Delivering Renewable Energy Under Devolution: Initial Findings Summary Report* (DREUD, 2013); G. Ellis, R. Cowell, F. Sherry-Brennan, P. Strachan, D. Toke, 'Planning, Energy and Devolution in the UK' 84(3) *Town Planning Review* 397 (2013); D. Toke, F. Sherry-Brennan, R. Cowell, G. Ellis, P. Strachan, 'Scotland, Renewable Energy and the Independence Debate: Will Head or Heart Rule the Roost?' 84(1) (January–March) *Political Quarterly* 61 (2013).

[156] These key devolved governmental actors include most importantly the UK's devolved governments and parliaments; see further Chapter 3.

[157] Muinzer, supra, n. 153 and as developed with Geraint Ellis of DREUD in supra, n. 14, 'Subnational Governance for the Low Carbon Energy Transition: Mapping the UK's "Energy Constitution"'.

researchers. Taking the devolution arrangements dimension, for example, this is in essence a constitutional issue, and it is the case that the UK's constitutional framework is broad and extensive. In addition to research challenges posed by the scale and scope of these circumstances, the constitutional framing itself is also highly complex. These intrinsic qualities pose difficult challenges for researchers engaged in resolving the sorts of problems that have just been highlighted.[158]

As things currently stand, and taking the energy sector as the site of primary research focus, Geraint Ellis of the DREUD team and the author have mapped and explored the UK constitution's allocation of energy-specific decarbonisation powers across the UK's 'multi-levels' of governance, honing in on the relationship between the 'national level' dominated by UK Government and the 'subnational level' dominated by the Devolved Administrations and their associated governance apparatuses.[159] In a paper entitled 'Subnational Governance for the Low Carbon Transition: Mapping the UK's "Energy Constitution"' (2017), an 'Energy Constitution' theoretical paradigm is developed.[160] This theoretical frame can be widely applied to energy-related governance matters, and, where applied more narrowly to the CCA in order to assess subnational governmental capacities in the context of energy decarbonisation, it insists that analysis of the pertinent elements of that framework is also to be explicitly fused with integrated assessment of energy-specific constitutional capacities for devolved action and constraint.[161] A theoretical outline and tabulated data are provided in the 'Energy Constitution' paper in order to provide an analytical template for the latter purpose.[162] This research can be combined usefully with previous research that has investigated the manner and extent to which the CCA itself accounts for devolved governance across its actual provisions in the context of the UK's national-subnational multilevels, published under the title 'Does the Climate Change Act 2008 Adequately Account for the UK's Devolved Jurisdictions?' (2016).[163]

---

[158] The national-subnational dimensions of the CCA are explored further in Chapter 3.

[159] Supra, n. 14, 'Subnational Governance for the Low Carbon Energy Transition: Mapping the UK's "Energy Constitution"'.

[160] Muinzer and Ellis, ibid.

[161] See Muinzer and Ellis, ibid., pp. 5–11, 'The UK's "Energy Constitution": Mapping the Powers'.

[162] See Muinzer and Ellis, ibid., pp. 5–11 and the Appendix, at pp. 19–22.

[163] Muinzer, supra, n. 147.

The substance of the important subnational dimensions of the CCA, which tends to be eclipsed by the 'national' features and processes embedded in the framework, is explored in more detail in Chapter 3. For present purposes, however, some general points may be drawn. It is clear that the subnational sphere *does* have a significant role to play in working towards the attainment of the CCA's national objectives. Broadly speaking, the UK's devolved jurisdictions enjoy significant political and legal space to strive progressively towards climate mitigation and adaptation outcomes, and this includes a degree of constitutional space that permits innovative and pragmatic governance action. On the other hand, the UK's devolved governments also have the facility to absorb the national climate regime's legally binding requirements in a largely passive way, adhering to and applying their national obligations in a chiefly automatic way where the Devolved Administrations are in effect merely 'doing what they are told' by the national level under the terms of the framework. In addition to the two major scenarios that have just been summarised as falling within the permitted parameters of the CCA—pragmatic, active subnational governance versus obligatory, passive subnational governance—a further broad possibility exists for any or all of the Devolved Administrations to use their constitutional space to be as obstructive as possible, that is to say, to work against the national level decarbonisation regime so far as is possible within the permitted constraints imposed by the CCA and the limits imposed by constitutional capacities for action.[164]

It is important to note that some of the issues raised in this area may be highly specific to the UK; for example, many countries do not have an equivalent or comparable devolved governance structure. This said, however, a range of concerns are at issue that do seem to point the way towards broader, generalisable truths, as follows:

---

[164]On the general attitudes and approaches that have been adopted by the UK's devolved territories, and their impact on the CCA's intended outcomes, see further Chapter 3.

- the CCA is a 'national' framework, where action is led at the national level but where it is also simultaneously applied 'downward' to subnational spheres of governance;
- the UK experience demonstrates that the subnational arena has a significant capacity to impact the success, or otherwise, of the national framework's outcomes;
- thus, a 'national' framework of this sort must be careful to adequately account for subnational and localized scales of governance, because a national regime of this type *will* be steadily impacted over time to some extent by the subnational arena, even though that impact may be somewhat hidden or eclipsed by the more overt 'national' features and orientation of the framework and the perceptions attaching to it;
- these points serve to emphasise that where other states seek to apply a CCA-style model to climate governance, the mechanisms underpinning the CCA will need to be adapted to fit the specific concrete multi-level governmental, legal (etc.) circumstances in which they are to be applied

One way to take the UK model and *improve* on it where a country is seeking to adopt it will be to calculate the likely impact of the subnational governance arena on the proposed framework as carefully and as closely as possible at the Bill stage or earlier, and factor this likely impact carefully into the form and nature of the final outcome Act.[165]

*

The next chapter will build on this chapter's treatment of the background to the CCA, the debate and commentary surrounding the framework, and the brief introduction to its substantive and procedural features, by examining and exploring the actual content of the CCA in detail.

---

[165] Questions arising around the most appropriate level(s) at which climate and energy initiatives should be spatially scaled can pose complex problems, and as such should be considered carefully in the context of any national climate regime design, see further: M.H. Benson, 'Regional Initiatives: Scaling the Climate Response and Responding to Conceptions of Scale' 100(4) *Annals of the Association of American Geographers* 1025 (2010); B.K. Sovacool, M.A. Brown, 'Scaling the Policy Response to Climate Change' 27(4) *Policy and Society* 317 (2009); M.R. Pasimeni, et al. 'Scales, Strategies and Actions for Effective Energy Planning: A Review' 65 *Energy Policy* 165 (2014); G. Bridge, et al. 'Geographies of Energy Transition: Space, Place and the Low Carbon Economy' 53 *Energy Policy* 331 (2013).

CHAPTER 2

The Content of the Act

Abstract This chapter examines the content of the Climate Change Act, doing so in a way that makes its complex legislative design clear and intelligible for the reader. It outlines and explains the framework's key requirements, obligations and procedures, and drills into aspects of particular interest or complexity. While the commentary may be drawn on to useful effect by lawyers seeking to familiarise themselves with the CCA, a primary objective is also to provide the first detailed general understanding of the content and design of the CCA for non-lawyers.

Keywords Content of the Climate Change Act 2008 · Environmental law and policy · Environmental legislation

1

This opening section briefly introduces the general content of the CCA and flags up and discusses the CCA's relatively *minor* dimensions. This clears the ground for the subsequent section (Chapter 2, Part 2), which engages in detail with the CCA's major components and innovations, and moves through the framework from beginning to end. In doing so, it focuses on distinguishing and exploring the CCA's *major* mechanisms and most important substantial innovations.

© The Author(s) 2019
T. L. Muinzer, *Climate and Energy Governance for the UK Low Carbon Transition*, https://doi.org/10.1007/978-3-319-94670-2_2

The CCA begins with its title, stated as 'Climate Change Act 2008', and this is followed with a short introductory paragraph, which is a conventional design feature of UK legislation. The short paragraph briefly summarises the key changes and additions to the law that the CCA has been designed to create. This opening portion is followed by the main body of the CCA, which is divided into a number of Parts, and these are followed in turn by a series of Schedules appended to the end of the Act. The title and short introductory section, the main Parts, and the Schedules that follow, amount to the full Act when taken together as a whole. Setting aside the title and short introductory paragraph, the contents of the CCA fall as follows:

Part 1 'Carbon Target and Budgeting'
Part 2 'The Committee on Climate Change'
Part 3 'Trading Schemes'
Part 4 'Impact of and Adaptation to Climate Change'
Part 5 'Other Provisions'
Part 6 'General Supplementary Provisions'

SCHEDULE 1 'The Committee on Climate Change'
SCHEDULE 2 'Trading Schemes'
SCHEDULE 3 'Trading Schemes Regulations: Further Provisions'
SCHEDULE 4 'Trading Schemes: Powers to Require Information'
SCHEDULE 5 'Waste Reduction Schemes'
SCHEDULE 6 'Charges for Carrier Bags'
SCHEDULE 7 'Renewable Transport Fuel Obligations'
SCHEDULE 8 'Carbon Emissions Reduction Targets'

As with any substantial item of legislation, some elements of the framework will be of major importance while others will be comparatively minor, and indeed, some elements may have virtually negligible importance insofar as they may involve somewhat incidental but technically necessary legal formalities. While the Act as a whole is undeniably important, in terms of the outline of the CCA's fundamental *substantive* and *procedural* duties and associated components that has been given in the preceding chapter, its most important elements are as follows:

Part 1 'Carbon Target and Budgeting'
Part 2 'The Committee on Climate Change'
(in conjunction with Schedule 1, 'The Committee on Climate Change')

Part 3 'Trading Schemes'
(in conjunction with Schedules 2-4, pertaining to the form and governance of trading schemes)
Part 4 'Impact of and Adaptation to Climate Change'

On the one hand, on its arrival the framework created fairly sweeping changes to UK law. On the other hand, it has also created certain changes of a narrower or more subsidiary nature, which have impacted relatively discrete areas of environmental law and governance in a manner intended to support the more sweeping changes underpinning the broader governance regime. The most prominent of the major sweeping changes to UK law that have been occasioned by the CCA have been highlighted in the preceding chapter. It has been seen that these include the creation of greenhouse gas emissions reduction targets for 2020 and 2050, and that a carbon budgeting system has been established in order to drive this transition. The CCA has also opened up powers that permit trading schemes to be created that can directly limit greenhouse gas emissions or otherwise encourage emissions-reducing activities. Further, the framework has established the CCC—the Committee on Climate Change, introduced in the preceding chapter—and created obligations pertaining to adaptation to climate change.

Minor Changes

In terms of narrower/more subsidiary changes to environmental law and governance created under the terms of the Act, these include the variation of powers permitting adjustments to be made to schemes that use financial incentives to reduce levels of domestic waste production and stimulate an increase in recycling. These adjustments were listed under the heading 'Waste Reduction Schemes' in the original CCA,[1] but they have since been embedded in the Environmental Protection Act 1990 and repealed (i.e. removed) from the CCA. This is chiefly because these schemes were not created by the CCA; rather, the CCA was adjusting aspects of a regime that already existed in this area. Originally, the CCA asserted that:

---

[1] See CCA, ss.71–75 (as originally enacted). See also CCA, s.76, 'Collection of household waste'. These original sections were targeted at England and Wales, rather than Northern Ireland and Scotland.

The purpose of a waste reduction scheme is to provide a financial incentive—
  (a) to produce less domestic waste, and
  (b) to recycle more of what is produced,

and accordingly to reduce the amount of residual domestic waste.

A waste reduction scheme—
  (a) may cover the whole or any part of the area of a waste collection authority, and
  (b) may apply to all domestic premises, to domestic premises other than those of a specified description or to specified descriptions of domestic premises.[2]

These elements of the CCA, and other elements relating to waste reduction schemes, were removed by the Localism Act 2011.[3]

In terms of the tangible text of the CCA itself, these repeals have cut sections 71–75 out of the CCA, and gutted Schedule 5 to the Act, such that an up-to-date version of the framework is devoid of text at these sections (in the main body) and paragraph entries (at Schedule 5).[4] The CCA waste reduction initiatives also include another modest adjustment to procedures concerning the collection of household waste that *is* still embedded in the CCA at the time of writing (i.e. it has not been repealed/removed); however, this change is itself applied through a change that the CCA makes to the Environmental Protection Act 1990; specifically, section 76 of the CCA amends section 46 of the Environmental Protection Act 1990.[5] This modest technical change asserts that a waste collection authority is not obliged to collect household waste that is not placed for collection in appropriate receptacles as part of a mandatory waste reduction scheme, or where an associated requirement is imposed by a waste collection authority.[6]

---

[2] CCA (as originally enacted), Schedule 5 Para 2 (quoting changes made by this Schedule to the Environmental Protection Act 1990).

[3] See Localism Act 2011, ss.47(a)–(b), 240(1)(e), Schedule 25 Part 8.

[4] The main body of a UK Act is conventionally comprised of 'sections', whereas Schedules that appear at the end of legislation are commonly composed of numbered 'paragraphs'.

[5] See further CCA, s.76; Environmental Protection Act 1990, s.46(11).

[6] See further ibid. (CCA, s.76, amending the Environmental Protection Act 1990, s.46; these changes apply to England and Wales only).

Powers are also opened up under the terms of the CCA that permit charges to be applied for single use carrier bags, namely the sorts of bags that shops commonly issue to people to allow them to carry their shopping. The powers are opened at section 77, and a framework for the governance and regulation of bag charges is fleshed out at Schedule 6 (and including the facility to apply a penalty of up to £5000 where bag charge regulations are breached[7]). Further narrow/subsidiary legal changes and innovations occasioned by the CCA include the creation of a power to alter some existing aspects of the UK Renewable Transport Fuel Obligation (hereafter 'RTFO').[8] UK Government has summarised the RTFO as follows:

> The [RTFO] supports the government's policy on reducing greenhouse gas emissions from vehicles by encouraging the production of biofuels that don't damage the environment. Under the RTFO suppliers of transport and non road mobile machinery (NRMM) fuel in the UK must be able to show that a percentage of the fuel they supply comes from renewable and sustainable sources. Fuel suppliers who supply at least 450,000 litres of fuel a year are affected. This includes suppliers of biofuels as well as suppliers of fossil fuel. The RTFO only covers biofuels used in the transport and NRMM sectors. …The 450,000 litre figure includes all fossil fuels and biofuels.[9]

A year prior to the CCA's creation, powers existing under the Energy Act 2004 had been used by UK Government to issue the RTFO Order 2007,[10] which established an RTFO scheme with an initial obligation period commencing in April 2008. The 2007 Order also created a non-departmental public body called the Office of the Renewable Fuels Agency in order to oversee the scheme's operation.[11] As such, and in a fashion resembling the CCA's waste reduction scheme components, the features of the CCA concerning RTFOs should not be misinterpreted as

---

[7] CCA, Schedule 6 Para 10(4).

[8] CCA, s.78, Schedule 7.

[9] DfT, *Guidance: Renewable Transport Fuels Obligation* (HM Government, 2012), unpaginated version. See further 'The Renewable Transport Fuel Obligation', pp. 5–6, in DfT, *RTFO Guidance Part One, Process Guidance* (HM Government, 2017).

[10] Renewable Transport Fuel Obligations Order 2007, S.I. 2007 No. 3072.

[11] UK Government has since closed the Agency down, absorbing its functions into the Department for Transport in 2011.

suggesting that the RTFO is a CCA-specific innovation, insofar as this has been a pre-standing regime enabled already under the terms of existing UK legislation (in this case, the Energy Act 2004).[12] In other words, the CCA merely makes certain adjustments to this existing regime, doing so in an intertextual way, where the CCA inserts text into the Energy Act 2004, such that the environmental lawyer seeking to interpret the RTFO framework is in effect referred back to the amended Energy Act 2004 and offspring legislation including the 2007 Order (as amended) that has just been mentioned. Thus, it is fair to say that the CCA is more incidental than central to the RTFO mechanism. This can be contrasted, for example, with UK carbon budgeting (examined below), where one must turn to the CCA itself for the authoritative regime, that is, the parent Act that sets the regime out and stipulates its form, nature and legally binding foundational components.

Similarly, the CCA affected changes to UK law enabling an increase in fines and penalties that can be imposed on conviction of certain pollution offences or where breach of pollution regulations has occurred. Again, the pollution penalisation framework has a sophisticated pre-existing design under UK environmental law, and as such, the CCA has made small tweaks and adjustments in this area, where the CCA is impacting that pre-existing framework but is substantially more incidental than central to it. This is reflected in the intertextual way in which environmental law behaves: the CCA applied these changes at CCA section 88,[13] but it did not create the changes by establishing its own regime; rather, it did so by amending the Clean Neighbourhoods and Environment Act 2005,[14] and the Environmental Permitting (England and Wales) Regulations 2007, so that the CCA changes were incorporated into pre-existing law.[15] Indeed, this broader framework has since been adjusted further in its own right by the widening of

[12] Energy Act 2004, Part 2, Chapter 5, ss.124–132, 'Renewable Transport Fuel Obligations'.

[13] CCA, s.88(1)–(2).

[14] See Clean Neighbourhoods and Environment Act 2005, s.105(2). The Pollution Prevention and Control Act 1999 permits fines on summary conviction of an offence, and these changes increased the quantum of the fines that can be applied.

[15] Environmental Permitting (England and Wales) Regulations 2007. The change helps to make applicable penalties comparable with penalties arising under the Waste Management Licensing Regulations 1994.

elements of the integration of environmental permitting and compliance systems (in England and Wales): see generally the Environmental Permitting (England and Wales) Regulations 2010.[16]

Other notable but relatively minor changes to environmental law and governance under the terms of the CCA include some adjustments to existing energy law in order to permit UK Government to impose certain narrowly drawn carbon emissions reduction obligations on particular designated parties, including electricity generators.[17] They also include an obligation placed on a government Minister to report annually to UK Parliament on progress made towards improving the energy efficiency and environmental sustainability levels of 'buildings that are part of the civil estate'.[18] While some (including the author) may find that the term 'civil estate' feels somewhat vague, the CCA clarifies that:

> a building is part of the civil estate if it is—
>
> (a) used for the purposes of central government administration, and
> (b) of a description of buildings for which, at the passing of this Act, the Treasury has responsibilities in relation to efficiency and sustainability.[19]

Other noteworthy minor or miscellaneous features of the CCA are highlighted in due course below.[20] As noted above, the purpose of this opening section to this chapter has been to provide introductory remarks on the general content of the CCA and to identify and discuss briefly the CCA's relatively minor dimensions. In the next section, the unfolding outline will move through the CCA's Parts and Schedules from beginning to end. In doing so, it will focus on distinguishing and exploring

---

[16] Environmental Permitting (England and Wales) Regulations 2010. Over the course of these changes, the 2010 Regulations repealed CCA, s.88(2); see further Schedule 28, Regulation 109 of the 2010 Regulations. After being amended 15 times in their own right, the 2010 Regulations have since been consolidated under the Environmental Permitting (England and Wales) Regulations 2016.

[17] See further 'Carbon Emissions Reduction Targets', below, within section 2 under the heading 'Part 5 "Other Provisions"'.

[18] Quoting CCA, s.86(1).

[19] CCA, s.86(6)(a)–(b).

[20] See most particularly 'Part 5 "Other Provisions"' and 'Part 6 "General Supplementary Provisions"' below, under heading 2.

the CCA's major mechanisms and substantial innovations. The minor sorts of components outlined above may be treated or acknowledged as or where they arise, but the chief focus of interest will remain with the CCA's most important components.

2

The following analysis, which focuses most particularly on interpreting and understanding the most profound changes to UK environmental law and governance occasioned by the CCA, moves through the framework from beginning to end. It begins with CCA Part 1 and works through to Part 6. The Schedules that are appended to the Act are referenced by and interact with these Parts, and so the Schedules are treated as and where appropriate over the course of the Part 1–6 analysis.

PART 1 'CARBON TARGET AND BUDGETING'

In this Part, the CCA engages prominently with 'greenhouse gases'; thus it is important to be clear and precise about the intended meaning of this term in the context of the framework. It is stated that 'greenhouse gas' means 'any of the following':

(a) carbon dioxide (CO_2),
(b) methane (CH_4),
(c) nitrous oxide (N_2O),
(d) hydrofluorocarbons (HFCs),
(e) perfluorocarbons (PFCs),
(f) sulphur hexafluoride (SF_6).[21]

These gases are also described by the framework as 'targeted greenhouse gases'.[22] While this in effect amounts to a fixed definition, a degree of flexibility is also incorporated to the extent that the Secretary of State is expressly permitted to amend the 'targeted greenhouse gases' criteria.[23] Official amendment of the general meaning of 'greenhouse gas' is also permitted; however, the Secretary of State may only 'add to the gases listed in that definition' if another gas becomes recognised 'at

[21] CCA, s.92(1)(a)–(f).
[22] See CCA, s.24(1).
[23] CCA, s.24(1)(g), s.24(2).

[the] European or international level' as a pertinent contributor to climate change.[24] The 'targeted' greenhouse gases refer to the gases that are covered by the CCA's substantive reduction targets and budgets.[25] Elsewhere in the CCA, it is clarified that the word 'emissions' itself, 'in relation to a greenhouse gas, means emissions of that gas into the atmosphere that are attributable to human activity'.[26] In other words, the CCA is explicitly concerned with anthropogenic emissions.

The framework uses the indicator 'net UK emissions' as a description of the UK's total emissions for a given period, which is calculated by taking the total amount of targeted greenhouse gases actively released by the UK and then deducting removals of gases (removals occur through land use and forestry practices).[27] Part 1 of the CCA builds a sophisticated 'carbon budgeting' regime around these cumulative UK emissions outputs and in doing so establishes a 'net UK carbon account'.[28] The Part 1 carbon accounting scheme enables the creation of a finite number of 'carbon units', where each unit represents a standardised amount of greenhouse gas.[29] By gauging appropriate carbon budgeting levels from the 'net UK emissions' benchmark that has just been mentioned, fixed quantities of carbon units can be established that can be credited and debited to a UK carbon account by the Secretary of State over set periods of time. Thus, this process operates within the parameters of carefully calculated emissions limits and forms the basis of the CCA's practical carbon accounting mechanism; it means that limits on the level of permissible greenhouse gas outputs can be applied over set periods of time, such that a stipulated net UK carbon account level cannot be permissibly exceeded over a specified number of years. Since applicable carbon limits can be steadily reduced over time, this serves to drive

---

[24] CCA, s.92(2)–(3).

[25] See CCA, ss.24–25, 'Targeted greenhouse gases'; CCA, s.1, s.27(1).

[26] CCA, s.97. Greenhouse gas emissions are 'measured or calculated in tonnes of carbon dioxide equivalent'; CCA 2008, s.93(1).

[27] CCA, s.29(1)(a)–(c).

[28] CCA, ss.4–10, ss.26–28, s.1(1), s.4(1)(a). While the emphasis falls on securing the CCA's targets via domestic reductions, the UK can also acquire international credits in order to ease the domestic burden, with a limit being placed on the proportion of credits that can be of international origin in a given budgetary period. The Secretary of State is accorded powers to place limits on carbon unit use at CCA, s.11.

[29] The term 'carbon unit' is defined at CCA, s.26(1).

down emissions progressively. The CCA measures out its carbon budget periods in 5-year blocks of time, described as 'budgetary periods', and the initial period ran over 2008–2012.[30] Taken generally, this carbon accounting process is a complicated, sophisticated scheme, but one that has achieved significant success to date in practice.

Section 5(4) of the CCA had originally stipulated that *only carbon dioxide emissions* were to be used to set the carbon budget cap in the drive towards 2020. This section was repealed subsequently, meaning that when one reads the up-to-date CCA a blank space appears where the restriction of incorporated gases to carbon dioxide for these purposes had been stipulated.[31] The change was made by the Secretary of State, who used authority granted under section 6(4) to repeal section 5(4), which was due to occur under the terms of the CCA when the framework's 2020 target was being revised; it has been noted in the previous chapter under heading 2 that the CCA interim reduction target had initially been set at 26% before being revised upward to 34%, a change applied by the Secretary of State upon receiving the approval of UK Parliament. The change was made by issuing an Order, entitled the Climate Change Act 2008 (2020 Target, Credit Limit and Definitions) Order 2009.[32]

In terms of actually setting the 5-year carbon budgets, the budget limits are established by the Secretary of State well in advance. The CCA states that '[t]he carbon budget for a budgetary period... must be set for the periods 2008-2012, 2013-2017 and 2018-2022, before 1st June 2009', and 'for any later period', the budget must be set 'not later than 30th June in the 12th year before the beginning of the period in question'.[33] Similar time limits are placed on the Secretary of State's facility to set the amounts of carbon units.[34] While the carbon budgets are spread over 5-year periods, it was the case prior to the CCA's enactment that some had thought that it would be much wiser to set

[30] CCA, s.4(1)(a). Article 3(1) set a 55,000,000 carbon units limit for the net UK carbon account covering 2018–2022.

[31] See CCA, s.5(4). The change was made by the Climate Change Act 2008 (2020 Target, Credit Limit and Definitions) Order 2009, Arts 1 and 2(3).

[32] Climate Change Act 2008 (2020 Target, Credit Limit and Definitions) Order 2009 (S.I. 2009/1258).

[33] CCA, s.4(2)(a)–(b).

[34] CCA, s.11.

budget-style targets annually, i.e., for each year instead of over 5-year blocks. However, this would have compromised the flexibility of the 5-year approach, where the Secretary of State in effect has 5 years to 'get things right', such that where emissions reduction performance for one year in a 5-year cycle may prove weaker or more erratic than anticipated, there is a capacity to correct this by compensating for the shortfall across other budget years. The annual approach may also have occasioned a slightly heavier administrative burden, arising from a need to engage in pronounced major review, calibration and reporting exercises with a more frequent intensity, that is to say, by employing intense and rapidly occurring 1-year milestones in order to sketch out the general working trajectory rather than the 5-year milestones. While this sort of potentially heavier burden might be self-justifying, it might also be the case that the excess administrative energy it could possibly drain could be utilised more effectively in other ways.[35]

Nevertheless, the CCA also in some sense facilitates the best of both worlds, in that it permits for the 5-year budgets to be broken down into conceptual 'annual equivalents', as follows: 'The "annual equivalent", in relation to the carbon budget for a period, means the amount of the carbon budget for the period divided by the number of years in the period'.[36] More importantly, however, the CCA builds in some broader flexibility permitting amounts from one budget period to be carried into another. The rules on this flexible capacity are stated in the following terms:

- The Secretary of State may decide to carry back part of the carbon budget for a budgetary period to the preceding budgetary period.
- The carbon budget for the later period is reduced, and that for the earlier period increased, by the amount carried back.
- The amount carried back… must not exceed 1% of the carbon budget for the later period.
- The Secretary of State may decide to carry forward the whole or part of any amount by which the carbon budget for a budgetary period exceeds the net UK carbon account for the period.

---

[35] This said, it is to be noted that Scotland has opted for an annual reporting regime, which has operated relatively successfully to date; see further Chapter 3.

[36] CCA, s.5(2).

- The amount of the carbon budget for the next budgetary period is increased by the amount carried forward.
- Any such decision must be made no later than 31st May in the second year after the end of the earlier of the two budgetary periods affected.[37]

This process of what the CCA describes as involving the 'carrying forward' and 'carrying back' of carbon budget amounts is often called 'banking' and 'borrowing' in environmental policy parlance.[38] If the carbon budget *is* exceeded for a set period, the Secretary of State must '[a]s soon as is reasonably practicable' lay before UK Parliament a report containing 'proposals and policies to compensate in future periods for the excess emissions'.[39]

Orders issued by the Secretary of State that set carbon budgets cannot be revoked after the point at which the budget was to be set by has passed[40]; however, if there have since been 'significant changes affecting the basis on which the previous decision was made', then such decisions can be amended in limited circumstances by the Secretary of State,[41] as long as the period that the budget applies to has not since ended.[42] The duration of budgetary periods, including their start and end dates, can be changed by the Secretary of State in limited circumstances under affirmative resolution procedure,[43] meaning that a draft of the item of law that the Secretary of State intends to use to make a change must receive the approval of both Houses of Parliament before the pertinent alteration is permitted.[44] Such changes cannot be made for general reasons, but rather need to be made where the Secretary of State deems the change(s) necessary in order to keep UK budgeting aligned with relevant developments at the international or European level.[45]

---

[37] Drawn from CCA, s.17(1)–(5).

[38] See, for example, DECC's *Impact Assessment for the Level of the Fifth Carbon Budget* (HM Government, 2016). This and similar documents frequently refer to 'banking' and 'borrowing' practices.

[39] CCA, s.19(1).

[40] CCA, s.21(1).

[41] CCA, s.21(2).

[42] CCA, s.21(4).

[43] CCA, s.23(1)(a)–(b), s.23(6).

[44] 'Both houses of Parliament' indicate the House of Commons and House of Lords that comprise UK Parliament. See further CCA, s.91.

[45] CCA, s.23(2).

Section 26 defines what a 'carbon unit' is:

1. In this Part a 'carbon unit' means a unit of a kind specified in regulations made by the Secretary of State and representing—

   (a) a reduction in an amount of greenhouse gas emissions,
   (b) the removal of an amount of greenhouse gas from the atmosphere, or
   (c) an amount of greenhouse gas emissions allowed under a scheme or arrangement imposing a limit on such emissions.[46]

CCA Part 1 gives the Secretary of State extensive powers to regulate carbon units.[47] Section 27 defines the 'net UK Carbon account' and asserts that the Secretary of State must create regulations to govern the workings of the account, including in particular the crediting and debiting of carbon units to the account. The 'net UK carbon account' is defined in the following terms:

...the "net UK carbon account" for a period means the amount of net UK emissions of targeted greenhouse gases for the period—

   (a) reduced by the amount of carbon units credited to the net UK carbon account for the period in accordance with regulations under this section, and
   (b) increased by the amount of carbon units that in accordance with such regulations are to be debited from the net UK carbon account for the period.[48]

It is added that '[t]he net amount of carbon units credited to the net UK carbon account for a budgetary period must not exceed the limit set' on the use of carbon units for any given budget period.[49] Sections 26 and 27 of the CCA, as qualified by section 28, sketch out a range of powers that permit the Secretary of State to create and issue regulations in order to administer the scheme effectively. These elements of the CCA are light on detail concerning the substantive form and content of the regulations themselves that might flow from these powers, being largely

[46] CCA, s.26(1)(a)–(c).
[47] CCA, s.26(2)–(4).
[48] CCA, s.27(1)(a)–(b).
[49] CCA, s.27(2). See also CCA, s.11, entitled 'Limit of use on carbon units'.

intended to simply open up a flexible regulation-creating capacity, rather than setting a rigid framework to be incorporated by the Secretary of State into the actual regulations that result.[50]

Taken generally, the scheme is concerned with establishing, monitoring and reducing an overall reading of UK emissions. Thus, it is important that the CCA also establishes the intended meaning of 'net UK emissions' measurements, as follows:

(a) "UK emissions", in relation to a greenhouse gas, means emissions of that gas from sources in the United Kingdom;
(b) "UK removals", in relation to a greenhouse gas, means removals of that gas from the atmosphere due to land use, land-use change or forestry activities in the United Kingdom;
(c) the *"net UK emissions" for a period*, in relation to a greenhouse gas, *means the amount of UK emissions of that gas for the period reduced by the amount for the period of UK removals of that gas.*[51]

It is also added that '[t]he amount of UK emissions and UK removals of a greenhouse gas for a period must be determined consistently with international carbon reporting practice'.[52] In principle, that practice is dictated by the United Nations Framework Convention on Climate Change (UNFCCC) and the agreements that have flowed from it, including the Kyoto Protocol and the Paris Agreement.[53]

Certain matters must be taken into account by the Secretary of State in relation to any decisions taken under Part 1 concerning carbon budgets, including the setting of budget levels, and also by the CCC where that committee is issuing advice relating to a decision.[54] The matters to be taken into account include:

– energy policy, and in particular the likely impact of the decision on energy supplies and the carbon and energy intensity of the economy.[55]

The other matters to be taken into account are:

[50] CCA, s.26(2)–(4), s.27(3)–(5), s.28.
[51] CCA, 29(1)(a)–(c). Emphasis added.
[52] CCA, s.29(2).
[53] See further Chapter 3.
[54] On the CCC, see further below.
[55] CCA, s.10(2)(f).

- scientific knowledge about climate change
- technology relevant to climate change
- economic circumstances, and in particular the likely impact of the decision on the economy and the competitiveness of particular sectors of the economy
- fiscal circumstances, and in particular the likely impact of the decision on taxation, public spending and public borrowing
- social circumstances, and in particular the likely impact of the decision on fuel poverty
- differences in circumstances between England, Wales, Scotland and Northern Ireland
- circumstances at European and international level
- the estimated amount of reportable emissions from international aviation and international shipping for the budgetary period or periods in question.[56]

In sum, under the general terms of CCA Part 1, the Secretary of State must establish 'carbon budgets' covering set 5-year intervals. The carbon budgets equate to UK emissions levels over those 5-year periods, and 'carbon units' facilitate a process of carbon accounting, where the units can be credited or debited to the UK's net carbon account. The Secretary of State is obligated to report on the measures the government will take to achieve the carbon budget requirements.[57] He is also obligated to report on the measures that will be taken to correct matters through compensation in future budget periods in the event that a carbon budget limit has been exceeded.[58] UK emissions levels and general progress under the scheme must also be reported to UK Parliament by the Secretary of State.

In conjunction with the features raised above, Part 1 also creates and imposes the CCA's 2020 and 2050 emissions reduction targets. As noted at several points throughout this book, the CCA applies an 80% emissions reduction target based on 1990 emissions levels for 2050 and a 34% emissions reduction target based on the 1990 baseline levels for 2020. The CCA gives the '1990 baseline' itself a slightly flexible meaning, insofar as this baseline *does* refer to the net UK emissions of carbon dioxide for 1990,[59] but it *also* refers to the net UK emissions for

---

[56] CCA, s.10(2)(a)–(e) and (g)–(i).
[57] CCA, s.14.
[58] CCA, s.19.
[59] CCA s.1(2)(a).

the 'base year' pertinent to the framework's other 'targeted' greenhouse gases, some of which differ from 1990 in practice.[60] Thus, as just noted, the carbon dioxide 'base year' is 1990, and although the 1990 baseline nominally refers to the general baseline emissions marker for the targeted gases, section 25(1) indicates that the base years for some of the other targeted greenhouse gases differ in practice. The base years for these gases run as follows:

methane, 1990
nitrous oxide, 1990
hydrofluorocarbons, 1995
perfluorocarbons, 1995
sulphur hexafluoride, 1995[61]

The Secretary of State is imbued with power to issue orders altering these base years (and to set base years for any new targeted greenhouse gases that may be added over time), but only under limited circumstances, namely where there are 'developments in European or international law or policy that make it appropriate to do so'.[62] Thus, for example, economic constraints do not entitle the Secretary of State to relax the base years; the general implication under the framework is that European and international developments could require the Secretary of State to tighten the regime (though in principle such powers could also be used to reduce it). The CCA's utilisation of a 1990 baseline marker has been influenced by international developments, where 1990 is the baseline year employed under the international-level Kyoto Protocol for carbon dioxide emissions.[63] As noted above, under section 24 of the CCA, and subject to various qualifications and procedures, the Secretary of State can add more targeted greenhouse gases to the framework's carbon accounting regime if developments in the future suggest that the list should be drawn more widely.

The 2050 and 2020 targets themselves appear in Part 1 at section 1(1) and section 5(1)(a) and are phrased in the following terms:

---

[60] CCA, s.1(2)(b).
[61] CCA, s.25(1).
[62] CCA, s.25(4).
[63] The baseline is set by Article 4(2)(b) of the UNFCCC.

1 The target for 2050

> (1) It is the duty of the Secretary of State to ensure that the net UK carbon account for the year 2050 is at least 80% lower than the 1990 baseline.

...

5 Level of carbon budgets

> (1) The carbon budget—
>
>> (a) for the budgetary period including the year 2020, must be such that the annual equivalent of the carbon budget for the period is at least 34% lower than the 1990 baseline[.][64]

Clearly, these targets are framed in the language of the CCA's carbon accounting regime, which serves to emphasise that the attainment of the targets embodies the scheme's most immediate and important set of overall headline outcomes. It is also notable that the duty to achieve the targets here, and the primary duties assigned elsewhere across the CCA, normally fall to the 'Secretary of State'. In the UK, 'Secretary of State' is a title that refers to a Cabinet Minister in charge of a UK Government department.[65] Thus, it appears within the CCA as a term referring to the pertinent Cabinet Minister from UK Government who is leading the government department relevant to the issues at hand. In this case, the Minister at the head of UK Government's Department of Energy and Climate Change (DECC) had amounted to the primary 'Secretary of State' being indicated by the CCA; however, DECC was controversially abolished in July 2016 by the then new Prime Minister Theresa May.[66] DECC's functions have since been rolled into UK Government's

---

[64] CCA, s.1(1), s.5(1)(a).

[65] Strictly speaking, under the terms of UK constitutional law there is technically a single office of Secretary of State; however, in practice responsibility is delegated to individual Secretary of States (as noted in the main text). See further A.J. Simcock, 'One and Many— The Office of Secretary of State' 70(4) *Public Administration* 535 (1992). The author is grateful to Prof Richard Macrory QC for drawing attention to this source over the course of research.

[66] See D. Castelvecchi, 'New Brexit Government Spells Shake-Up for Science: Theresa May Promotes a Former Science Minister and Abolishes Climate-Change Department' 535(7612) *Nature* 331 (2016). While DECC took the lead on mitigation, the Department for Environment, Food and Rural Affairs also had some key responsibilities in the area of adaptation.

Department for Business, Energy and Industrial Strategy (BEIS). As such, the 'Secretary of State' designated with primary duties under the CCA currently indicates the Minister leading BEIS, namely the Secretary of State for Business, Energy and Industrial Strategy.

These emissions reduction duties and the carbon budgeting system that enmeshes them have thus far proven effective in driving down UK emissions in a consistent and steady way. In spite of their success at the present time, it should also be noted that the previous chapter in this book has highlighted that the CCA's headline substantive duties have an uncommon character that poses some problems for environmental lawyers and courts. Most particularly, some uncertainty exists around the extent to which the legally binding target duties can be meaningfully enforced.[67] While Part 1 of the CCA frames the targets as being outcomes that must be achieved by the Secretary of State in *absolute* terms, there is an implicit sense that the Secretary of State must *endeavour* to achieve the outcomes as best he can, which is not entirely the same thing; as such, it might be fair to say that in practice the targets are 'aspirational' rather than 'absolute'.[68] Reid has stressed that the new-form substantive outcomes required here by the CCA do not simply place a duty on the Secretary of State to achieve something that is attainable in some sort of isolation, but rather require the Secretary of State to *steer the cumulative conduct of a wide range of governance actors* as best he can in order to achieve the specified ends.[69] There will be substantial limits on the Secretary of State's available steering capacity in practice. For instance,

[67] See further the discussion of these issues in Chapter 1. It is notable that the Northern Irish courts have drawn on the CCA over the course of wrestling with tricky issues evoked by statutory duties centring on appropriate residential accommodation being made available by pertinent authorities to a hospital patient suffering from a learning disability, see: *In the Matter of an Application By JR 47 for Judicial Review* [2013] NIQB 7. The CCA was evoked over the course of addressing concerns surrounding the way in which the courts ought to approach complex issues underpinning statutory duties: see further McCloskey J.'s comments at ibid., Para [35]. In his learned judgement, McCloskey J. draws attention to the fact that Lord Hope has stressed in an earlier case entitled *R (G) v Barnett LBC* ([2004] 2 AC 208) that one of the central features of target duties is that they are 'concerned with general principles and not designed to confer absolute rights on individuals'; see ibid., Para [35].

[68] See, e.g., D. Feldman, 'Legislation Which Bears No Law' 37(3) *Statute Law Review* 212 (2016).

[69] C.T. Reid, 'A New Sort of Duty? The Significance of "Outcome" Duties in the Climate Change and Child Poverty Acts' *Public Law* 749 (2012), p. 749.

powerful economic and finance controls are concentrated most directly in the HM Treasury department, rather than the Department for Business, Energy and Industrial Strategy. Bearing this in mind, for a 'hopeless' endeavour by claimants to persuade the courts that the CCA created a particular legitimate expectation in the context of UK Treasury's handling of investment in the Royal Bank of Scotland, see the court case *R. (on the application of People & Planet) v HM Treasury*.[70] Further, in the case *R. (on the application of Friends of the Earth) v Secretary of State for Energy & Climate Change*[71] it was held by the courts that the Secretary of State *could* take account of budgetary constraints in considering the reasonable practicability of implementing a fuel poverty strategy, and although the main piece of legislation at issue was a different Act of Parliament,[72] the CCA itself was evoked over the course of legal argument.[73] Thus, Christopher Forsythe's comment that the CCA targets may be interpreted as embodying 'an aspiration' rather than 'a guarantee of achievement', while debatable, is perhaps unsurprising.[74]

A further feature of note in relation to CCA Part 1 concerns the extent to which the framework is calibrated to interact with broader environmental policy. It is common for UK Government to establish policies and then to subsequently create law in order to implement those policies; indeed, this has occurred in the case of the CCA in the sense that the CCA is an important legal manifestation of crucial aspects of the UK's Low Carbon Transition policy process. However, the framework itself also acts as a direct and active launch pad for additional comprehensive policy activity. This is particularly evident around sections 13–15, where

[70] *R. (on the application of People & Planet) v HM Treasury* [2009] EWHC 3020 (Admin); quoting Mr Justice Sales, ibid., Para [11].

[71] *R. (on the application of Friends of the Earth) v Secretary of State for Energy & Climate Change* [2009] EWCA Civ 810.

[72] The Warm Homes and Energy Conservation Act 2000, s.2. Note, however, that in this case the duty in the legislation was qualified by the term 'as far as reasonably practicable', whereas no such qualification appears in the CCA, thus arguably giving the CCA duties a significantly harder edge.

[73] See *R. (on the application of Friends of the Earth)*, supra, n. 71, Para [18] of Lord Justice Maurice Kay's lead judgement.

[74] Joint Committee on the Draft Climate Change Bill, *Oral and Written Evidence (second report); Draft Climate Change Bill* (2006-7, HL 170-II, HC 542-II) 239–240 (Evidence of Christopher Forsythe to the Joint Committee); see also J. Church, 'Enforcing the Climate Change Act' 4(1) *UCL Journal of Law and Jurisprudence* 109 (2015), p. 116.

it is asserted that the 'Secretary of State must prepare such proposals and policies as the Secretary of State considers will enable the carbon budgets that have been set under this Act to be met'.[75] The strategies put forward are subject to a long-term imperative, in that the 'proposals and policies must be prepared with a view to meeting the [2050 emissions reduction] target in section 1' and 'targets [set] for later years'.[76]

In practice, it is notable that Part 1 of the CCA, and indeed the entire CCA in general, has not gained much explicit traction in court litigation and the case law of the UK courts to date. Thus far the framework's carbon target and budgeting system has tended towards having been drawn on as a flexible instrument forming a part of overall legal debate, as for instance in the case *Preston New Road Action Group v Secretary of State for Communities and Local Government*, involving an appeal against planning permission granted for exploration works to test the feasibility of developing hydraulic fracturing/'fracking' at two sites in Lancashire, England.[77] Here, Lord Justice Lindblom gave fairly short shrift to the argument that opening the door to exploration works in this way would conflict in law with the aims of the CCA by likely leading to a detrimental increase in greenhouse gas emissions:

> The idea that, in a project of exploration for shale gas such as this, as opposed to the commercial production of shale gas, the substitution of new gas for existing gas in the grid will raise the total consumption of gas by increasing gas usage, that significant additional greenhouse gas emissions are thus likely, and that there might be some conflict with the objectives of the Climate Change Act 2008, gains no credence in the report of the Committee on Climate Change, "Onshore Petroleum: The compatibility of onshore petroleum with meeting the UK's carbon budgets", published in March 2016, or in the Government's response, published in July 2016. [This does] not serve to demonstrate that such consequences are likely.[78]

Nevertheless, it is fair to say that the CCA's broad architecture and pervasive nature is permitting it to crystallise gradually more strongly in general legal discourse and in judicial and court considerations and

---

[75] CCA, s.13(1).

[76] CCA, s.13(2)(a)–(b).

[77] *Preston New Road Action Group v Secretary of State for Communities and Local Government Court of Appeal* (Civil Division) [2018] EWCA Civ 9.

[78] *Preston New Road Action Group v Secretary of State*, ibid., Para [72] (Lord Justice Lindblom).

reasoning; see, for instance, recognition of its general governance role in *R. (on the application of Drax Power Ltd) v HM Treasury*[79] over the course of Mr Justice Day's treatment of an unsuccessful challenge to UK Government's rather sudden removal of important financial exemptions for renewable source electricity.[80]

A rare example to date of the CCA being accorded a substantial determinative influence in a significant court ruling is provided by *The Queen on the Application of London Borough of Hillingdon & Ors v Secretary of State for Transport v Transport for London*.[81] Here, the CCA was drawn upon in order to challenge UK Government's favourable disposition towards the development of a controversial third runway at Heathrow Airport. The Secretary of State had indicated in a White Paper strategy that UK Government's support for a third runway at Heathrow was dependant on certain climate change (and other) conditions being satisfactory, and then subsequently indicated that those conditions could be met. However, the court found the Secretary of State had erred, insofar as the CCA had been passed since that strategy was set out and thus the position would need to be reviewed in the light of how those developments would impact the conditions. Thus, Lord Justice Carnwath asserted that 'common sense demanded that a policy established in 2003, before the important developments in climate change policy, symbolised by the Climate Change Act 2008, should be subject to review in the light of those developments'.[82] In addition to CCA section 1, CCA section 32 figured significantly in the case, a section pertaining to the CCC; the CCC had offered general advice over the course of its duties to the effect that aviation emissions should not increase beyond a certain limit, and the role of this advice was factored influentially into the legal argument and reasoning.[83]

---

[79] *R. (on the application of Drax Power Ltd) v HM Treasury* [2016] EWHC 228 (Admin).

[80] *R. (on the application of Drax Power Ltd)*, ibid., Para [12]. See also *Solar Century Holdings Limited & Others v Secretary of State for Energy & Climate Change* [2014] EWHC 3677 (Admin), Para [16] (Mr Justice Green).

[81] *The Queen on the Application of London Borough of Hillingdon & Ors v Secretary of State for Transport v Transport for London* [2010] EWHC 626 (Admin).

[82] *The Queen on the Application of London Borough of Hillingdon & Ors*, ibid., Para [52].

[83] For detail, see the court ruling itself, and see further the discussion in R. Macrory *Regulation, Enforcement and Governance in Environmental Law* (2nd edition, Hart, UK, 2014), pp. 270–271.

The most significant instance of action in the courts—indeed, the only major action to date that expressly hinges on major elements of the CCA in a pronounced way—is underway at the time of writing. The challenge concerns *Plan B and Others* (Claimants) v *Secretary of State for Business, Energy and Industrial Strategy* (Defendant), where, at the time of writing, the Claimants are seeking permission from the High Court of Justice to proceed with a judicial review case in England. On 8 December 2017, the Claimants filed their grounds for judicial review at the High Court, that is to say, they submitted the necessary procedural documents containing the legal points that they wish to rely on as the basis of their challenge in court. Here, in their statement of the grounds and facts at issue, they asserted that, although the CCA's 80% reduction target for 2050 may have been roughly consistent with an existing intention to limit average global warming to 2 degrees C above pre-industrial levels over the CCA's initial and early stages, the Paris Agreement has since been ratified. This subsequent international agreement (discussed in Chapter 3 of this book) asserts that the general intention is now to go further by endeavouring to limit average global warming to 1.5 degrees C above pre-industrial levels. Given that the UK is a signatory to the Paris Agreement, the Claimants have argued that these developments in international law and scientific understanding dictate that the Secretary of State has an obligation under section 2 of the CCA to revise the 2050 80% reduction target upwards in the light of these new circumstances.[84] In conjunction with this, the Claimants have argued that the Secretary of State's failure to adjust the target is irrational and frustrates the legislative purpose of the CCA.[85] The Claimants are seeking declaratory relief that the Secretary of State has violated his responsibilities under the CCA, and they are also seeking a 'mandatory order that the Secretary of State revise the 2050 target',[86]

---

[84] Section 3 of the Judicial Review Claim Form that was filed by the claimants at the High Court summarises this matter as follows: 'Details of the decision to be judicially reviewed[:]... The ongoing failure of the Secretary of State for Business, Energy and Industrial Strategy not to exercise his power under section 2 of the Climate Change Act 2008 to amend the percentage figure set out in section 1(1) of that Act'.

[85] 'Irrationality'/'unreasonableness' amounts to a special technical category of law that can be evoked in the UK in order to judicially review the behaviour of public actors.

[86] Quoting section 7 of the Judicial Review Claim Form, which the author obtained from the courts service.

meaning they hope that the court will actively order the Secretary of State to adjust the 80% target. They are also seeking any additional relief that the court might consider to be appropriate and wish to be awarded costs in order to cover their legal expenses.[87]

PART 2 'THE COMMITTEE ON CLIMATE CHANGE'

Part 2 establishes the UK's CCC. The CCA designs the CCC to function as an expert advisory body that can advise, examine and report on the emissions reduction regime rolled out under the framework. It is a statutory non-departmental public body, and so it retains some meaningful degree of independence from UK Government.[88] Important elements of the CCC's main advisory functions are drawn together under CCA sections 33–35.[89] Although the CCC liaises and works in a targeted advisory capacity with each of the Devolved Administrations of Northern Ireland, Scotland and Wales, in general terms its duties are drawn in a UK-wide manner; in other words, the CCC's advisory functions cover the UK as a whole.

---

[87] The Claimants have presented five grounds to the High Court in seeking judicial review of the Secretary of State's purported failure to revise the 2050 target: (1) it is ultra vires, because it frustrates the legislative purpose of the CCA; (2) it is based on an error of law regarding the objective of the Paris Agreement; (3) it is irrational, because it fails to take into account and / or inappropriately weighs considerations including the scientific risks of global climate change and developments in international law, most notably in relation to the Paris Agreement; (4) it violates the Human Rights Act 1998; (5) it breaches the public sector equality duty set out in Section 149 of the Equality Act 2010. Concerning point (4), relating to human rights, it is claimed that: under the Human Rights Act 1998, Article 2 ('Right to Life'), Article 8 ('Right to Respect for Private and Family Life'), and Article 1 of Protocol 1 ('Protection of Property') have been breached individually in their own right and jointly when read with Article 14 (concerning 'Prohibition of Discrimination'). [Immediately prior to this book going to print, *Plan B's* application has been refused by Mr Justice Supperstone at the High Court, see: *R (Plan B Earth and Others) v. Secretary of State for Business Energy and Industrial Strategy (Defendant) and the Committee on Climate Change (Interested Party)* [2018] EWHC 1892 (Admin). The author has been informed by *Plan B* that it will appeal this decision.]

[88] CCA, Schedule 1 Para 27. It is to be noted that there is no general power granted to government Ministers to direct the CCC, which contrasts with certain other operational non-departmental public bodies, including the Environment Agency (the primary body responsible for environmental regulation and protection in England).

[89] CCA, ss.33–35.

The CCC has a duty to advise the Secretary of State where he is considering bringing forward major substantive changes to the decarbonisation regime, including where amendment to the 2050 target is being considered[90] and where carbon budgets are being set.[91] In addition to its advisory capacities,[92] the CCC is also possessed of important reporting powers and duties, where it must report to UK Parliament. Most particularly, the CCC provides an extensive report on emissions reduction progress and general overall performance under the terms of the carbon accounting regime, which is laid before UK Parliament each year.[93] This report is also laid before the devolved parliaments.[94] Over the course of the CCC's overarching reporting processes, the Secretary of State and the Devolved Administrations can issue the CCC with certain guidance and directions that it will take into account.[95] The overall importance of the role played by the CCC in terms of scrutinising progress, advising and reporting is highly significant and should not be underestimated. Thus, for example, Lockwood, who has praised the innovative character of the CCA, has noted in particular that without the CCA's long-term legal targets working in conjunction with the CCC standing behind them, it is quite likely that a carbon budget for the 2020s would not have been agreed to by UK Government.[96] Fankhauser, Averchenkova and Finnegan have described the CCC as the 'custodian of UK climate policy'[97] and 'the fulcrum of the UK climate change architecture'.[98]

The CCC enjoys a significant range of freedom in carrying out its functions. Thus, it is permitted to:

---

[90] CCA, s.33.

[91] CCA, s.34.

[92] The CCC's advisory obligations have been widened slightly since the creation of the CCA: see the Infrastructure Act 2015, s.49, which places an obligation on the CCC in its own right to advise the government on the impact of onshore petroleum policy choices on UK decarbonisation.

[93] CCA, s.36(1).

[94] CCA, s.36(1).

[95] See CCA, ss.41–42.

[96] M. Lockwood, 'The Political Sustainability of Climate Policy: The Case of the UK Climate Change Act', 23(5) Global Environmental Change 1339 (2013), p. 1346.

[97] S. Fankhauser, A. Averchenkova, J. Finnegan, *10 Years of the UK Climate Change Act* (Grantham Research Institute and London School of Economics, 2018). p. 2.

[98] Fankhauser et al., supra n. 52, p. 5.

- enter into contracts,
- acquire, hold and dispose of property,
- borrow money,
- accept gifts, and
- invest money.[99]

In addition to gathering information and carrying out research and analysis, it may also 'commission others to carry out such activities'[100] and publish the results of such work as appropriate.[101]

In addition to the CCC's mandatory active advisory and reporting duties, it must also remain generally amenable to providing advice *on request* where a national authority seeks its assistance. Thus, the CCA requires that:

The Committee must, at the request of a national authority, provide advice, analysis, information or other assistance to the authority in connection with—

(a) the authority's functions under this Act,
(b) the progress made towards meeting the objectives set by or under this Act,
(c) adaptation to climate change, or
(d) any other matter relating to climate change.[102]

'Particular' attention is drawn to the fact that the CCC must be amenable on request to advising national authorities about 'any limit proposed to be set by a trading scheme on the total amount of the activities to which the scheme applies'[103] and to 'assist the authority in connection with the preparation of statistics relating to greenhouse gas emissions'.[104]

Schedule 1, which appears towards the end of the CCA, sets out detail on the CCC's structure, staff, practices and general administration. The CCC panel members are appointed jointly by the Secretary of State, the Scottish Ministers, the Welsh Ministers and the relevant Northern

---

[99] CCA, s.39(2)(a)–(e).

[100] CCA, s.39(3)(b).

[101] CCA, s.39(3)(c).

[102] CCA, s.38(1)(a)–(d).

[103] CCA, s.38(2)(a). See also CCA, s.48 on the CCC's duty to provide advice on trading scheme regulations. Trading schemes are discussed further below.

[104] CCA, s.38(2)(b).

Ireland department (i.e. the 'national authorities').[105] If the Secretary of State wishes to amend the number of panel members, then the consent of the other national authorities is to be sought.[106] In addition to establishing the CCC itself, the CCA has also created a subcommittee of the CCC, called the Adaptation Sub-Committee (hereafter 'ASC'). This has been achieved most particularly by CCA Schedule 1 Para 16: here, it is stated at Paragraph 16(1) that the ASC is to be created, and the legislation then proceeds to set out aspects of the form, nature and functions of the ASC.[107] The ASC continues to do significant work in the UK at the present time. The CCA asserts that 'The ASC must provide the [CCC] with such advice, analysis, information or other assistance as the [CCC] may require in connection with the exercise of its functions' in relation to matters concerning climate change adaptation.[108] Further, it is notable that the CCA gives the CCC an *option* to establish additional sub-committees,[109] however note that the ASC itself has been *mandatorily* established as a subcommittee by the legislation.

In general terms, it is fair to say that the CCA creates conditions where a strong sensitivity and degree of value is accorded to the opinions and policy advice of the CCC; however, the CCA at no point imbues the CCC with any active degree of power that can permit it to trump or overrule the ascendant authority of the Secretary of State. A further aspect of particular note, which might amount to a significant weakness in the framework, concerns the observation that the CCC's advisory and reporting functions might be said credibly to engage the issue of *mitigation* significantly more fulsomely than the issue of *adaptation*; however, this is perhaps of little surprise, insofar as this trend in effect follows the form of the CCA itself, which is most prominently concerned with mitigation. That said, the less fulsomely treated issue of adaptation does feature importantly in the framework (see the discussion of CCA Part 4 below), and in addition to the counterbalance provided by the presence of the ASC, it is also the case that certain special duties are placed on the

---

[105] See CCA, Schedule 1 Paras 1–2.

[106] CCA, Schedule 1 Paras 1(4)–1(5).

[107] CCA, Schedule 1 Para 16, 'The Adaptation Sub-Committee'.

[108] CCA, Schedule 1 Para 16(10).

[109] CCA, Schedule 1 Para 15.

CCC to advise on climate change impact[110] and to report on progress in connection with adaptation.[111]

Although the provisions of the CCA have yet to see much meaningful action in the UK courts, certain elements of the framework have featured notably in R. *(on the application of Griffin) v Newham LBC Divisional Court.*[112] These have included provisions concerning the CCC's advisory role. The case involved a local authority's decision to vary a planning permission so as to enable a greater number of flights per year at London City Airport, which was challenged (in part) on grounds relating to the CCA in the courts. It was argued that a statement from the Transport Secretary in UK Government issued in January 2009 relating to a UK target for aviation emissions could be viewed as creating a limit on increased capacity at airports. Here, the Minister had indicated that a new target was to be established for aviation emissions and had sought policy advice from the CCC as part of this process. The Secretary of State had confirmed the imposition of a new target in April 2009. It was argued before the High Court that pertinent Ministerial statements amounted to a fundamental change in aviation and climate change policy, such that the local authority had erred in discharging its legal obligations by failing to factor that change into its decision concerning expanding the number of airport flights. In its ruling, the court held that these matters did *not* amount to a relevant policy change that should have caused the planning committee to change its approach. In practice, that type of approach, which was rejected by the court, would have involved the planning committee departing from a White Paper that was functioning as the appropriate policy statement in its decision-making processes. The case raised points around the CCA's sections 1, 13 and 35, which deal, respectively, with the 2050 target (section 1), the Secretary of State's duty to prepare proposals and policies for meeting carbon budgets (section 13), and the CCC's advisory competence relating to international aviation and shipping emissions (section 35).[113]

[110] CCA, s.57.

[111] CCA, s.59.

[112] R. *(on the application of Griffin) v Newham LBC Divisional Court* [2011] EWHC 53 (Admin).

[113] See most particularly R. *(on the application of Griffin)*, ibid., Para [31] of Lord Justice Pill's lead judgement, and Para [38] dismissing this element of the application.

PART 3 'TRADING SCHEMES'

Part 3 of the CCA enables trading schemes to be created. Emissions trading schemes can provide an effective means of making a significant contribution to greenhouse gas mitigation, especially where reduction of emissions sourced from energy generation is concerned.[114] The capacity to create regulations for trading schemes is accorded to 'national authorities', who are granted special powers under Part 3 to issue secondary legislation for this purpose. Elsewhere, at CCA section 95, the general meaning of 'national authority' is stated as follows:

> 1. In this Act 'national authority' means any of the following—
> (a) the Secretary of State;
> (b) the Scottish Ministers;
> (c) the Welsh Ministers;
> (d) the relevant Northern Ireland department.[115]

Within Part 3 itself, section 47 in effect emphasises that these are the relevant 'national authorities' engaged directly by the CCA's trading scheme framework.[116] In other words, the national authorities imbued with direct powers to create and regulate trading schemes are pertinent governmental actors in UK Government (i.e. the Secretary of State) and the devolved governments of Scotland, Wales and Northern Ireland. As noted below, however, these powers can be used to create/appoint special administrators, such that the practical burden of administrating trading schemes does not have to be borne directly by the national authorities themselves.

A 'trading scheme' is defined by the CCA in rather vague language,[117] as follows:

---

[114]Consider, for example, the substantial effect that the EU's Emissions Trading Scheme ('ETS') has had in driving down emissions from energy generation across the EU. See further: Council Directive 2009/29/EC [2009] OJ L140/63 (the ETS Directive).

[115]CCA, s.95(1)(a)–(d).

[116]CCA, s.47.

[117]The definition is unhelpfully vague insofar as many sorts of schemes may, for example, operate by 'limiting or encouraging the limitation of activities that consist of the emission of greenhouse gas' (point (a) of the definition quoted immediately below in the main text) without amounting to being trading schemes. For clarity on the form and nature of the sorts of trading schemes that tend to predominate in the context of climate and energy governance, see J. Robinson, J. Barton, C. Dodwell, M. Heydon, L. Milton, *Climate Change Law: Emissions Trading in the EU and the UK* (Cameron May, London, 2007).

A 'trading scheme' is a scheme that operates by–

(a) limiting or encouraging the limitation of activities that consist of the emission of greenhouse gas or that cause or contribute, directly or indirectly, to such emissions, or
(b) encouraging activities that consist of, or that cause or contribute, directly or indirectly, to reductions in greenhouse gas emissions or the removal of greenhouse gas from the atmosphere.[118]

Parts 1 and 2 of Schedule 2 to the CCA improve on the vagueness that prevails here by setting out a range of rules that underpin these schemes, and section 45 offers some guidance on how one should interpret the sorts of 'indirect' greenhouse gas emissions activities that are mentioned in the definition above.[119] Thus, for instance, if the burning of fossil fuels at energy plants can be said to be a clear example of a 'direct' contribution to such emissions, the 'consumption' of the generated energy by people in their homes arising as a consequence can be said to make an 'indirect' contribution.[120]

Schedule 2 Part 1 focuses on fleshing out certain regulatory criteria pertaining to trading schemes that serve to limit greenhouse gas emitting activities (in other words, it concerns trading scheme definition (a) above). Schedule 2 Part 2 focuses on trading schemes that encourage activities that can contribute to greenhouse gas emissions reduction (therefore engaging trading scheme definition (b) above). Part 3 of Schedule 2 sets down guidelines pertaining to the administration and enforcement of the schemes. In general terms, the type of scheme provided for at trading scheme definition (a) above will include the conventional 'cap and trade' style of emissions trading scheme that tends to predominate in the sphere of climate governance. These sorts of schemes place a 'cap' or limit on emissions and create carbon allowances with a market value up to the level of the cap. An active market is opened up, such that companies and other market actors responsible for emissions can trade allowances and credits within the cap.[121]

---

[118]CCA, s.44(2)(a)–(b). In addition to being vague (ibid.), the combined phrasing of the following elements at point (b) seems grammatically unsatisfactory: 'encouraging activities that consist of... to reductions in greenhouse gas emissions'.

[119]CCA, s.45.

[120]See CCA, s.45(1)(a).

[121]For further detail on these important schemes, see M. Faure, M. Peeters (eds.), *Climate Change and European Emissions Trading* (Edward Elgar, Cheltenham, 2008).

The trading scheme framework is UK-wide[122] (although it also permits the UK to fit in with the EU's current emissions trading scheme regime; on the EU, see further Chapter 3). The Devolved Administrations are also imbued with a targeted capacity to participate and regulate certain aspects within their jurisdictions (as noted above). This makes some good degree of constitutional sense where it is borne in mind that the Devolved Administrations have various devolved areas of governance that fall directly under their control rather than the control of the national level (and thus including the control of the Secretary of State). It could create severe governance obstacles if a Devolved Administration was trying to regulate some aspect of its devolved competences while at the same time the Secretary of State was planning unilaterally to apply some form of trading scheme in that area, thereby not only complicating matters but also potentially disrupting the Devolved Administration's facility for coherent governance in some way.[123] The CCA permits the Devolved Administrations to establish trading schemes in their areas of competence, that is to say, within the range of the powers that have been devolved to them by UK Parliament, but all other matters fall within the Secretary of State's trading scheme remit.[124] The manner in which the framework permits the Devolved Administrations to administrate and enforce aspects of trading schemes at the subnational level[125] harmonises with certain aspects of the CCA's broader approach to its carbon budgeting system. For instance, under the CCA Part 1 carbon budgeting regime, both (national) Ministers of the Crown and UK Government departments and (subnational) Scottish Ministers, Welsh Ministers and equivalent (ministerial) Northern Ireland departments are granted a facility to acquire and dispose of carbon units.[126]

Although certain primary powers are granted to national authorities in relation to the creation of trading schemes and the associated facility

---

[122] 'This Part applies to activities carried on in the United Kingdom, regardless of where the related emissions, reductions or removals of greenhouse gas occur', CCA, s.45(3).

[123] For targeted exploration of the relationship between national and devolved authorities within the parameters of the CCA, see further T.L. Muinzer, 'Does the Climate Change Act 2008 Adequately Account for the UK's Devolved Jurisdictions?' 25(3) *European Energy and Environmental Law Review* 87 (2016).

[124] CCA, s.47(7).

[125] See further CCA, Schedule 2 Para 21.3.b–d.

[126] See CCA, s.87, read in conjunction with ss.26–27.

to issue pertinent regulations, it does not follow that those authorities will necessarily actively administrate a given trading scheme in practice. This is due to the CCA enabling those authorities to appoint administrators for such purposes. Thus, it is asserted by the CCA that an administrator that is appointed to oversee a trading scheme(s) will be a public body that can be, but does not have to be, one of the national authorities.[127] Further, the CCA permits powers of enforcement to be accorded to national authorities/administrators under regulations issued for such purposes, meaning that trading scheme compliance can be enforced.[128] Civil financial penalties for non-compliance can be created and imposed under the regulations.[129] The regulations may also create criminal offences that can be punishable on summary conviction by a maximum of a one year term of imprisonment, or by a fine up to £50,000, or by both of these punishments.[130]

Spotlight on CCA Schedules 2–4

Within the CCA, after the main Parts of the CCA have been set out—including Part 3 covering 'trading schemes'—a range of fairly extensive Schedules appear at the end of the legislation. Schedules 2–4 contain substantial further detail on the Part 3 trading schemes. Schedule 2 in effect fleshes out the substance of the regulations that can be issued by national authorities under the CCA in this area. Schedule 2 Part 1 begins by addressing trading scheme definition (a) that has been noted above,[131] namely *schemes that limit activities*. Most particularly, it includes the stipulation that these sorts of regulations are to specify the pertinent trading scheme time periods, the nature of the activities that the scheme in question applies to, the units of measurement of the activities for the purposes of the scheme and the participants in the scheme.[132]

---

[127] See particularly CCA, Schedule 2 Para 21.3.e.

[128] CCA, Schedule 2 Para 28.

[129] CCA, Schedule 2 Para 29.

[130] See CCA, Schedule 2 Para 30. 'Summary' convictions, while sometimes involving very serious issues, generally involve less severe offences than 'indictments'; the CCA includes further provision for punishment on indictment at Schedule 2 Para 30.7.

[131] That is, the definition of a 'trading scheme' stated at CCA, s.44(2)(a).

[132] See CCA, Schedule 2 Paras 2–4.

Detail is also given on the allocation and use of allowances, where allowances represent 'the right to carry on a specified amount of the activities in a trading period'.[133] The regulations being made to enable trading schemes can also incorporate the use of credits, where credits represent '(a) a reduction in an amount of greenhouse gas emissions, or (b) the removal of an amount of greenhouse gas from the atmosphere'.[134] Provision is made for requiring payments to be made to the pertinent national authority or administrator in the event that a trading scheme participant falls short in traded allowances or credits.[135] The regulations can also require trading scheme participants to hold a special permit.[136] Further, they can establish that a CCA trading scheme can be blended/take part in other trading schemes to some extent, including trading schemes established 'at United Kingdom, European or international level'.[137] Thus, for instance, in principle an emissions trading scheme designed to act on UK energy generation can also be permitted to incorporate pertinent EU-level trading (the EU ETS[138]) or indeed international-level trading (namely, allowances and credits opened under a Kyoto Protocol-style mechanism[139]).

Schedule 2 Part 2 moves on to the sorts of schemes covered by trading scheme definition (b) above,[140] namely *schemes that encourage activities*. In general terms, this portion of the framework largely follows the same type of template for regulations laid out in Schedule 2 Part 1 (i.e. the template employed for definition (a) trading schemes), with the most important differences relating to the inclusion in this Part of the setting of targets and obligations[141] and the issue of 'certificates evidencing the carrying on of the activities in a trading period'.[142] On the first point,

---

[133] CCA, Schedule 2 Para 5(1).

[134] CCA, Schedule 2 Para 7(1)(a)–(b).

[135] CCA, Schedule 2 Para 8.

[136] CCA, Schedule 2 Para 10.

[137] CCA, Schedule 2 Para 11; quoting from Para 11(1)(b).

[138] That is, the European Union's Emissions Trading Scheme established under the Emissions Trading Scheme Directive (as noted earlier in Chapter 1 and also raised in Chapter 3).

[139] The Kyoto Protocol is discussed in Chapter 3.

[140] That is, the definition of a 'trading scheme' stated at CCA, s.44(2)(b).

[141] CCA, Schedule 2 Para 16.

[142] CCA, Schedule 2 Para 17(1).

targets and obligations, schemes under this Part must establish targets that are to be achieved by participants over trading periods. The certificates that are issued in effect take the place of the allowances and credits that predominate under the activity-limiting framework of Schedule 2 Part 1; these certificates are to be tradable,[143] and each certificate represents a measurement of the amount of emissions-reducing activity that has been carried out. It is notable that there appears to be no overt restriction placed by the CCA on blending aspects of schemes that limit activities (definition (a)) and schemes that encourage activities (definition (b)) within the regulations, such that trading schemes might be designed to treat both of these dimensions.

Schedule 3 lays down additional provisions relating to regulations made under the CCA Part 3 enabling powers. Schedule 3 Part 1 states the procedure to be followed where regulations are made by a single national authority, and Part 2 elaborates the procedure for issuing regulations that are made by more than one national authority.[144] Schedule 3 Part 3 contains technical legal information on the process of issuing trading scheme regulations via Orders in Council.[145] An Order in Council is a special order created by the government and approved symbolically by the person of the UK Monarch.[146] The device is given a role here for technical legal reasons in certain instances where a joint trading scheme will extend to Scotland, or to Scotland and one or more of Northern Ireland, Wales and England.[147]

Schedule 4 confers special 'powers to require information' upon relevant environmental authorities for the purposes of enabling trading schemes to be created.[148] For example, if a national authority was seeking to design a trading scheme that related to the supply of electricity, pertinent electricity suppliers could be compelled to provide the

---

[143] CCA, Schedule 2 Para 19(1).

[144] See generally CCA, Schedule 3 Parts 1 and 2.

[145] CCA, Schedule 3 Part 3, 'Power to Make Provision by Order in Council'.

[146] Per the Glossary on the official UK Parliament website: 'Orders in Council are used when an ordinary statutory instrument would be inappropriate, such as for transferring responsibilities between government departments. They are issued by and with the advice of HM Privy Council and are approved in person by the monarch'. http://www.parliament.uk/site-information/glossary/orders-in-council/.

[147] See CCA, Schedule 3 Para 9(2)(a)–(b).

[148] CCA, Schedule 4 (quoting from the title of the Schedule).

authority with necessary information for the purpose. It is to be noted, however, that the Schedule is only 6 Paragraphs long, and Paragraphs 1–5 now no longer apply.[149] This is because the paragraphs were designed specifically to permit swift information gathering from the period of the CCA's first enactment (in 2008) up to the end of 2010, in order to facilitate the creation of a particular trading scheme, called the Carbon Reduction Commitment. This was a targeted emissions reduction scheme that was to be mandatorily applied in the UK to large non-energy intensive public and private organisations. This mandatory scheme continues to run in the UK at the time of writing,[150] and it has produced some positive results.

Part 4 'Impact of and Adaptation to Climate Change'

As the title that introduces Part 4 of the CCA suggests—'Impact of and Adaptation to Climate Change'—this segment of the framework shifts the CCA's primary focus very much onto the issue of climate change *adaptation*, in contrast to the preceding Parts, which are concerned foremost with the issue of *mitigation*. CCA Part 4 places a duty on the Secretary of State to 'lay reports before Parliament containing an assessment of the risks for the United Kingdom of the current and predicted impact of climate change'.[151] It is stipulated that the first report is to be produced within three years of Part 4 of the CCA coming into force.[152] The CCA received Royal Assent on 26 November 2008, and CCA section 100, entitled 'Commencement', stipulates that Part 4 of the CCA 'come[s] into force at the end of two months beginning with the day it is passed'.[153] This points towards a deadline of 26 January 2009 (i.e. 2 months after 26 November 2008). In the UK Government fashion of often cutting things rather fine, the required adaptation report appeared

---

[149] These paragraphs were revoked by the CCA itself, at CCA, s.50(2), which placed a time-bar on Paras 1–5. Schedule 4 Para 6, which remains active law, concerns disclosure of information; it permits certain information relating to trading schemes to be shared where necessary amongst environmental authorities or a trading scheme's administrator.

[150] UK Government has signalled that it intends to close the scheme after the 2018–2019 compliance year.

[151] CCA, s.56(1).

[152] CCA, s.56(2).

[153] CCA, s.100(5).

on 25 January 2012, where it was laid before UK Parliament.[154] Subsequent mandatory reports are to be published in cycles of no later than five years after the report that has preceded it.[155] The CCC has a duty to advise on these reports.[156]

Part 4 also makes a distinction between the process of *reporting* on adaptation and the obligation to produce *programmes* for adaptation to climate change.[157] Thus, while the points just raised concern obligations to assess and report on impacts and developments in the sphere of adaptation, in addition to the Secretary of State's reporting functions section 58 also requires him to produce climate change adaptation programmes, as follows:

1. It is the duty of the Secretary of State to lay programmes before Parliament setting out

    (a) the objectives of Her Majesty's Government in the United Kingdom in relation to adaptation to climate change,
    (b) the Government's proposals and policies for meeting those objectives, and
    (c) the time-scales for introducing those proposals and policies[.][158]

A given programme must address the risks identified in the most recent adaptation report that has been issued at that time.[159] Further, the CCC is placed under its own free-standing duty to report to UK Parliament on the progress that is being made towards the implementation of the programmes.[160]

---

[154]DEFRA, *UK Climate Change Risk Assessment: Government Report* (HM Government, 2012).

[155]CCA, s.56(3). It is accepted that delays may arise in publishing reports, but that the Secretary of State must give reasons for any delay and specify when the report will be completed; CCA, s.56(4).

[156]CCA, s.57(1).

[157]CCA, s.58, 'Programme for Adaptation to Climate Change'.

[158]CCA, s.58(1)(a)–(c). In Northern Ireland, tailored regional adaptation programmes are to be produced along similar lines by Northern Ireland's Devolved Administration and brought before the Northern Ireland Assembly at regular intervals; CCA, s.60. Special criteria relating to aspects of reporting on adaptation in Wales in the context of devolved Welsh functions is also set out at CCA, ss.66–69.

[159]CCA, s.58(1).

[160]CCA, s.59.

The Devolved Administrations engage substantially in their own adaptation planning beneath the substate leve, but the Secretary of State's assessment and reporting functions treat and encompass the UK-wide adaptation experience, such that his remit incorporates detailed assessment of adaptation across the UK as a whole. Thus, it is unsurprising that significant input from and liaison with the Devolved Administrations is required, and including the assertion that final assessment (and programme) documents 'must' be sent to the Devolved Administrations.[161] The Secretary of State is given powers to direct pertinent bodies of a public nature to provide targeted reports on adaptation, and he can issue formal guidance on what is required in terms of the process and content underpinning what is reported back.[162] Certain blocks are put on the Secretary of State's facility to encroach in a direct way on devolved powers, that is to say, governance areas subject to some degree of substantial control by the Devolved Administrations. A consultation process with the Devolved Administration in question can be required where the Secretary of State seeks to use powers to issue guidance or directions that may raise issues to this end, and in certain cases, the devolved authority's active formal consent will be necessary.[163]

Part 5 'Other Provisions'

Part 5 creates or enables a variety of narrower measures that are designed to contribute to lowering UK greenhouse gas emissions. Some notable aspects of these features have been addressed above already, under heading 1 in this chapter. These measures chiefly concern: waste reduction schemes; household waste collection; charges for carrier bags; the RTFO (i.e. the Renewable Transport Fuel Obligation, discussed earlier). While the matters in Part 5 are significant, they tend to be much less of a major and innovative product of the CCA than the far-reaching mechanisms underlying CCA Parts 1–4 and their associated Schedules can be said to represent. This is reflected in the way that these features are treated within the CCA itself, where, with the exception of carrier bag charges, the CCA is largely making changes or tweaks to *pre-existing*

---

[161] This applies to both climate change impact reports (CCA, s.56(6)) and adaptation programmes (CCA, s.58(4)).

[162] CCA, ss.61–62.

[163] See CCA, s.64(1)–(2).

frameworks that *already* govern the area in question. Thus, the CCA's waste reduction scheme provisions make substantial adjustments to the Environmental Protection Act 1990, the collection of household waste elements also change that Act, and the RTFO provisions alter the Energy Act 2004.

While these elements of the CCA have also been discussed above under heading 1, such that the reader should refer to that section for additional comment, in the commentary to follow these provisions are briefly treated in the order in which they appear in Part 5 to the CCA.

Waste Reduction Schemes

Waste reduction schemes are covered in Part 5 of the CCA under sections 71–75 and Schedule 5; however, each of these sections and the Schedule have since been repealed by the Localism Act 2011, meaning they are no longer active law under the terms of the CCA itself. Consequently, where one looks at an up-to-date version of the CCA, empty gaps appear where the text of the original sections and Schedule 5 had appeared. As has been noted above, these elements of the CCA created changes to another piece of UK legislation, an important environmental statute called the Environmental Protection Act 1990: the CCA's original provisions asserted that changes were to be made to this 1990 Act, such that the CCA provisions enabling the creation of waste reduction schemes have been bolted into that pre-existing statute. Therefore, although the waste reduction scheme elements are now erased from the CCA, they are present in another framework and as such they are still active law.[164] The waste reduction schemes enabled under the terms of the Environmental Protection Act 1990 apply to England and Wales and are designed to target occupants of domestic premises, using economic mechanisms to encourage the production of less waste and more recycling. Regulations can be made for this purpose, and the economic elements of the scheme can be applied through mandatory payments or charges, and including through adjustments to council tax.[165]

---

[164] See most importantly Environmental Protection Act 1990, s.60A and Schedule 2AA, both entitled 'Waste reduction schemes'.

[165] See further Environmental Protection Act 1990, Schedule 2AA, where the rules are set out.

Collection of Household Waste

A small technical change is made by the CCA to the Environmental Protection Act 1990 in order to improve the latter Act's waste management regime. The change means that a waste collection authority is not obliged to collect household waste that is not placed for collection in appropriate receptacles in certain circumstances.[166]

Charges for Carrier Bags

The CCA opens up powers to place charges on single use carrier bags.[167] For example, this allows the government to rule that where a shopper is being given a plastic bag(s) at a shop in order to carry one's shopping, the person must be charged for that bag(s).[168] Further detail and rules for employing these powers are set out at Schedule 6.[169] Here, national authorities are given detailed powers to create regulations, appoint bag charge administrators, and generally ensure that carrier bag charging can operate in practice.

Renewable Transport Fuel Obligations ('RTFOs')

The CCA makes changes to the Energy Act 2004.[170] It has been noted above under heading 1 in this chapter that an RTFO scheme has been enabled under the Energy Act 2004 that places requirements on certain transport fuel suppliers to source a certain percentage of the supplied fuel from renewable sources. The CCA changes make some adjustments

---

[166]CCA, s.76, amending the Environmental Protection Act 1990, s.46. These changes apply to England and Wales.

[167]The carrier bag provisions here do not extend to Scotland; see CCA, s.99(3). Scotland has a levy of its own, however, applied by the Single Use Carrier Bags Charge (Scotland) Regulations 2014, which were issued under powers made available to the Scottish Ministers under ss.88, 89, 90 and 96(2) of the Climate Change (Scotland) Act 2009. Scotland's 2009 Act is considered in Chapter 3.

[168]CCA, s.77.

[169]CCA, s.77, Schedule 6.

[170]CCA, s.78, Schedule 7; Schedule 7 amends the Energy Act 2004, Chapter 5, Part 2.

to aspects of the Secretary of State's pre-existing powers[171] to regulate in this area, appoint administrative bodies, process monies accrued under the scheme, and associated matters.[172] These changes are bolted into the Energy Act 2004 by the CCA; as such, where the RTFO is concerned, it is the Energy Act that is the primary source of authority, and the CCA is more akin to a subsidiary agent of change.

Carbon Emissions Reduction Targets

Section 79 and Schedule 8 are entitled 'Carbon emissions reduction targets'.[173] These elements of the CCA make technical legal changes to other items of law so that they are compatible with the Secretary of State's powers to establish carbon emissions reduction targets and obligations within the decarbonisation parameters of the CCA regime.[174] They permit carbon emissions savings to be achieved through allowing for the imposition of targets and obligations on electricity generators, electricity distributors, electricity suppliers and other related parties. The amendments to existing law made by the CCA here cover the Gas Act 1986, the Electricity Act 1989 and the Utilities Act 2000.[175]

Miscellaneous

Part 5 closes by laying down some additional miscellaneous provisions. These include provisions in relation to reporting requirements in Wales, where the Welsh Devolved Administration is required to produce climate mitigation and adaptation reports tailored to Wales, and aspects of the ability to publish an energy measures report and guidance for Welsh local authorities are transferred from UK Government to the

---

[171] It is to be noted that the RTFO is rolled into the portfolio of UK Government's Department for Transport, and as such the pertinent Secretary of State indicated in these changes to the Energy Act 2004 is the Secretary of State for Transport. This differs from the 'Secretary of State' imbued with primary responsibilities under the CCA's main regime, namely the Secretary of State for Business, Energy and Industrial Strategy.

[172] See further CCA, Schedule 7, 'Renewable Transport Fuel Obligations'.

[173] CCA, s.79, Schedule 8.

[174] This mechanism does not apply to Northern Ireland; see CCA, s.99(4).

[175] See CCA, Schedule 8.

Welsh government.[176] The Secretary of State is also given a responsibility to issue mandatory guidance to companies and others on the matter of emissions reporting,[177] and a facility is created giving the Secretary of State an option to introduce mandatory emissions reporting by companies.[178] Further: the Minister for the Cabinet Office (a senior member of UK Government) is to report each year on how energy efficiency and sustainability is improving on the civil estate[179]; UK Government and the Devolved Administrations are given a capacity to acquire and dispose of carbon units and equivalent units in emissions trading schemes[180]; and some tweaks to existing law are made so that fines applied for committing a pollution offence can be raised.[181]

Part 6 'General Supplementary Provisions'

This final Part is less weighty in terms of substantial content than the preceding Parts of the CCA. It largely sets out some legal elements that clarify certain technical aspects of the CCA for lawyers. In particular, it clarifies the territorial reach of the CCA,[182] which spans the UK as a whole but that contains some narrower elements that only extend to part of the UK. For example, it has been highlighted above that aspects of adjustments to waste reduction schemes extend to England and Wales

---

[176] See further CCA, ss.80–82. The transfer of power between the national and Welsh authorities required amendment to the Climate Change and Sustainable Energy Act 2006. Elements relating to local authorities in Wales contained in CCA, s.81 are marked 'prospective', meaning they are not (yet) in law at the time of writing.

[177] See CCA, ss.83–84.

[178] CCA, s.85.

[179] CCA, s.86. See further the discussion of the civil estate above, in part 1 to this chapter.

[180] CCA, s.87. This mechanism permits the national authorities, namely UK Government and the Devolved Administrations, to offset greenhouse gas emissions by acquiring and disposing of carbon units directly.

[181] This required the CCA to amend the Clean Neighbourhoods and Environment Act 2005, s.105(2). The amendment is applied at CCA, s.88. The 2005 Act extends only to England and Wales.

[182] CCA, s.89, s.99.

only; this is clarified by section 99(2)(a).[183] Part 6 also stipulates certain technical legal elements relating to the processes underpinning the creation of the orders and regulations that can be made under the terms of the CCA,[184] and it provides a range of definitions that permit lawyers and others to interpret the technical terminology running through the framework.[185]

*

The next chapter will develop the unfolding understanding of the CCA by exploring its relationship to the international and subnational levels of governance. Consideration of the subnational experience will include reflection on the devolved territories of Northern Ireland, Scotland and Wales, and particular attention will be accorded to the Scottish Parliament's Climate Change (Scotland) Act 2009.

[183] CCA, s.99(2)(a).
[184] CCA, ss.90–91.
[185] CCA, ss.92–98.

CHAPTER 3

Multilevel Drivers: The International Level and the Devolved Level (Northern Ireland, Scotland and Wales)

Abstract This chapter considers the Climate Change Act from a 'multilevel' perspective, directly integrating the international level and the subnational level of governance into the analysis. It begins with consideration of the impact of the international level on the UK framework and also reverses this focus in order to recognise the influence that the Climate Change Act has exerted itself on the international arena. Next, the analysis proceeds to critique the 'national' Climate Change Act's relationship to the UK's complex subnational level, honing in on the UK's devolved jurisdictions (Northern Ireland, Scotland and Wales). Special detailed attention is accorded towards the end of the chapter to Scotland's pioneering Climate Change (Scotland) Act 2009.

Keywords International climate governance · Subnational climate governance · Climate Change (Scotland) Act 2009 · Climate law and policy in Northern Ireland, Scotland and Wales

1

This chapter commences with consideration of the international arena's influence on, and relationship to, the CCA, before moving on to address the UK's subnational governance setting. International developments have exerted a significant influence on the UK climate and energy governance experience. The most important initial agreement intended to tackle climate change in a comprehensive way at the international level can be traced to the United Nations Framework Convention on Climate Change (hereafter 'UNFCCC'), which was adopted by the United Nations in 1992.[1] This pioneering and for that time reasonably far-reaching international agreement provided a coherent legal framework that helped to formalise and structure an emerging international will to address the problem of anthropogenic global warming. While the UNFCCC did help to cohere a will towards unified international action, and included some significant concrete obligations,[2] it was generally very light on practical substance.[3]

In the late 1990s, the nature and substance of the action required in the context of the UNFCCC were developed substantially under the terms of the Kyoto Protocol. The Kyoto Protocol was set in place in 1997 and came into force in February 2005.[4] In terms of the overall cohort of international UNFCCC signatories, this important agreement honed in particularly on certain 'developed' countries listed in the UNFCCC's Annex I, which included the UK, and committed these more economically robust countries to a binding greenhouse gas reduction target of 'at least 5% below 1990 levels in the commitment period 2008 to 2012'.[5] In other words, the Kyoto Protocol applied a collective 5% greenhouse gas emissions reduction target based on 1990 levels to the applicable nations for the period 2008–2012. The framing of these requirements in 1997 helped to set the structural tone for the UK's subsequent CCA, which also employs an approach predicated on greenhouse gas percentage limits that are calibrated against 1990 levels

[1] The UNFCCC defines 'climate change' as: 'A change of climate which is attributed directly or indirectly to human activity that alters the composition of the global atmosphere and which is in addition to natural climate variability observed over comparable time periods'. UNFCCC, Article 1(2).

[2] For example, reporting duties on progress towards greenhouse gas emissions reduction (UNFCCC, Article 12).

[3] On the design of the UNFCCC, see further J. von Stein, 'The International Law and Politics of Climate Change: Ratification of the United Nations Framework Convention and the Kyoto Protocol' 52(2) *Journal of Conflict Resolution* 243 (2008).

[4] Kyoto Protocol to the United Nations Framework Convention on Climate Change.

[5] Kyoto Protocol, ibid., Article 3(1).

and pegged to clearly blocked-out time periods.[6] In taking this sort of line, the CCA is not merely absorbing aspects of an international model approach to the climate change challenge, but, more than that, it is employing a national version of elements of that model that speaks to a significant extent in concert with international action, and that can therefore harmonise with international environmental governance in this area with a useful degree of ease.

Although the Kyoto Protocol embodied a substantial degree of international-level progress that would help to inform the form and content of the CCA, its general successes should not be overstated. In particular, a number of key nations resisted calls to ratify it. Most importantly, the USA, China and India avoided participation in the binding emissions reduction drive.[7] The EU, which included the UK within it, was generally strongly supportive of the Kyoto Protocol, and did participate, but the absence of the USA, China and India meant that the other most abundant greenhouse gas emitters on the global stage undermined progress by their lack of direct participation. Some improvement occurred in 2002, as China and India did choose to subscribe to the agreement; however, they did not receive binding reduction targets. The USA persisted in refusing to ratify the Protocol.[8] International law in this area has some tendency towards uneven legal drafting, as exemplified by the manner in which headline Kyoto commitments were applied to the countries listed at Annex I to the UNFCCC, however this Annex I list included the USA within it even though the Kyoto obligations were not directly applied to the USA.[9] There were complex political reasons underpinning the USA's position here, arising from the country's administration signing the agreement in the knowledge that the Senate would be extremely

---

[6] See further Chapter 2, on the detail of the Act.

[7] An insightful account of the Kyoto Protocol in terms of both the negotiation process and the substance of the Protocol is provided by S. Oberthur, H.E. Ott, *The Kyoto Protocol: International Climate Policy for the 21st Century* (Springer, Germany, 1999).

[8] The USA continues to resist ratification at the time of writing; although the 2008–2012 Kyoto compliance period itself has closed, UNFCCC Article 22 keeps the option to subscribe to the Protocol open, such that compliance can be linked to future international action under the UNFCCC framework.

[9] See UNFCCC, Annex I.

unlikely to ratify it[10]; however, regardless of the political context, as a point of principle international environmental law should always be drafted in as clear and precise a way as possible.

As noted, at this time the EU incorporated the UK within it as a prominent and influential Member State. From inside the overall 5% reduction marker that was imposed on the developed nations for 2008–2012, the EU itself received an obligation to lower its collective greenhouse gas emissions by 8% from 1990 levels over that time bracket. Internally, the EU adopted a 'burden sharing' approach to its reduction commitments,[11] with the contribution that was to be made by the UK and the other EU Member States being calculated against the backdrop of each state's historic emissions and its perceived socio-economic capacity to meet what was deemed to be a roughly fair and proportional manageable share of the reduction burden. The UK received a 12.5% reduction for the 2008–2012 Kyoto time span, and this fed directly into the state's internal construction of the CCA in the lead-up to its enactment. In practice, the UK substantially exceeded its 12.5% baseline reduction target, hitting instead around 22%,[12] an over-achievement that would have been unlikely to occur without the augmented downward pressure on emissions levels created by the CCA.

For all its narrower influence within the UK in setting an important backdrop to the creation of the CCA, and for all its broader macro-success as the most progressive step forward in tackling the climate change challenge at the international level at that time, the meaningful impact of the Kyoto Protocol was undermined in part by the fact that its targets did not go far enough. There was also some sense that states' persistent capital-driven efforts to liberalise trade indicated a worrying lack of joined-up thinking between the sphere of international economic development and climate change mitigation and adaptation, a theme raised

---

[10] See J. Hovi, D.F Sprinz, G. Bang, 'Why the United States Did Not Become a Party to the Kyoto Protocol: German, Norwegian and US Perspectives' 18(1) *European Journal of International Relations* 129 (2010).

[11] In addition to being facilitated by the form and structure of the EU's own internal workings, this sort of approach was enabled by international provisions including Kyoto Protocol Article 4. See further M. Grubb, 'The Economics of the Kyoto Protocol' 4(3) *World Economics* 143 (2003).

[12] DECC, *UK Progress Towards GHG Emissions Reduction Targets: Statistical Release* (HM Government, 2015), p. 3, 'Kyoto Protocol Target'.

in the work of Charnovitz.[13] Further, and as noted, various particularly essential nations had refused to ratify the agreement, and Kyoto also left a question mark over what the course of international action should look like *after* the expiry of the 2008–2012 time period.

This latter problem—uncertainty around the form and nature of post-2012 action—was compounded by challenges concerning the negotiation of the next compliance phase, where the international community struggled greatly to reach a unifying agreement. As such, negotiations were heavily dragged out over a number of years.[14] Viewed in this international context, it is clear that the CCA very much put the UK on the front foot internally, insofar as the UK had crafted a sophisticated internal climate regime that was up and running from 2008 in a manner that spanned the first Kyoto Protocol compliance period (2008–2012), and that subsequently permitted the UK to extend meaningful climate mitigation and adaptation action immediately beyond this period, and onwards into the future. The locked-in, long-term design of the CCA allowed this progressive action to persist in spite of the at best destabilising and at worst detrimentally chaotic developments playing out internationally at that time.

In 2007, at COP 13[15] in Bali, Indonesia, the international community was able to draw together and agree on a number of decisions and plans, which it grouped under the collective title of the 'Bali Road Map'; amongst these documents, the Bali Action Plan was perhaps of particular importance, given the emphasis that it placed on practical action.[16] Reduced to its essentials, however, the Bali Road Map was in effect a framework pathway for negotiation and agreement that it was hoped could facilitate the creation of a new set of binding targets at COP 15, that is, the UN's Climate Change Conference hosted at Copenhagen in late 2009. This gathering did produce and approve a 'Copenhagen Accord',[17] which

---

[13] S. Charnovitz, 'Trade and Climate: Potential Conflicts and Synergies', in J. Aldy, et al. (eds.) *Beyond Kyoto: Advancing the International Effort Against Climate Change* (Pew Center on Global Climate Change, 2003).

[14] See Chapter 4, 'The Kyoto Protocol and Beyond', and subsequent chapters, in S. Afionis (ed.) *The European Union in International Climate Change Negotiations* (Routledge, UK, 2017).

[15] The 13th meeting of the Conference of the Parties (or 'COP') to the UNFCCC.

[16] The Bali Action Plan, Decision 1/CP.13.

[17] The Copenhagen Accord of 18 December 2009, Decision 2/CP.15 (United Nations, 2009).

explicitly recognised a common international desire to keep global temperatures below a dangerous 2 degrees Celsius/3.6 degrees Fahrenheit rise on pre-industrial levels,[18] but, disappointingly, no binding reduction targets were arrived at and applied in order to pave the way forward. Nor were any significantly robust action plans agreed and set in place that could sketch out a concrete trajectory for future binding international action.

Subsequent conferences at Cancun and Durban could not bring these outstanding problems to a concrete resolution, in spite of the fact that it was agreed at Durban that a new legally binding agreement was to be drawn up by 2015. This inability to extend the initial Kyoto arrangements in a meaningful, binding way came as a frustration to senior EU officials and diplomats, as the EU had been a major advocate of the initial Kyoto agreement and was pushing the international community to drive further agreement forward.[19] Both the UK's nested position within an EU that was comparatively progressive in the area of climate and energy governance and the nature of the UK's internal CCA regime itself served to shield the UK's climate mitigation and adaption processes to a significant extent from external governance influences that would have otherwise proven to be immensely disruptive.

At Doha in December 2012, the Kyoto parties agreed upon a 'Doha Amendment to the Kyoto Protocol'.[20] Here, the signatories finally established a second Kyoto commitment period, scheduled to run from 1 January 2013 to 31 December 2020. This second period provided for an 'overall commitment' to emissions cuts for the industrialised nations, which was set at 18% below 1990 levels.[21] This 2013–2020 compliance period, however, was initially framed as an indicative, largely aspirational phase that could be firmed up over time. Therefore, it cannot be viewed as an equivalent successor to the more robust 2008–2012 compliance period, due not only to its looser aspirational character but also to the relatively gaping time span between 2013 and 2020. More to the point,

[18] Copenhagen Accord, ibid., Article 1.

[19] Indeed, the EU had announced a commitment to ramp up an intended EU target of 20% greenhouse gas emissions reductions on 1990 levels for 2020 (discussed below) to a substantially higher 30% target if the UNFCCC developed countries would agree to take similarly progressive action; see, e.g., the European Commission's *Limiting Global Climate Change to 2 degrees Celsius: The Way Ahead for 2020 and Beyond*, COM (2007) 2 final, p. 2.

[20] Kyoto Protocol (Amendment) Decision 1/CMP.8.

[21] This required significant amendment to Kyoto Protocol Article 3.

the Doha Amendment has not yet come actively into force, because it will activate 90 days after three-quarters of the Kyoto Protocol parties formally deposit their instruments of acceptance, and this condition has not yet been met. This is highly unsatisfactory.[22]

In 2015 in Paris, a hope existed on the part of many that the international community would finally arrive at an agreed set of robust legally binding commitments that would be meaningfully applied, including what many hoped would be directly applicable nationally tailored reduction targets, comparable in important respects to the 2008–2012 compliance phase. After a hard negotiation phase characterised by a marked degree of difficulty in agreeing on collective action, the international community succeeded in producing the Paris Agreement.[23] Although the agreement's 'legal' character is soft in nature,[24] it has at least been ratified, and it went 'live' on 4 November 2016. Over the course of negotiation, the EU continued to position itself as a progressive voice endeavouring to drive the emissions reduction agreements forward, and as an EU Member State the UK was represented chiefly by EU-level diplomatic channels. While some have pointed out that the Paris Agreement arguably fails to live up to expectations,[25] it does require signatories to prevent dangerous and irreversible levels of climate change by holding the increase in the global average temperature below 2 degrees Celsius of warming above pre-industrial levels[26]; signatories are also to go a good deal further than this if possible, by simultaneously 'pursuing efforts' to limit the temperature increase to 1.5 degrees Celsius above pre-industrial levels.[27]

The general thrust is that the meaningful substance of the Paris Agreement will come into force in a practical way in or around 2020, such that the international governance regime will continue to run on seamlessly on the instant that the 2013–2020 Doha Amendment time

---

[22] See further P. Sands, J. Peel, with A. Fabra, R. MacKenzie, *Principles of International Environmental Law* (4th edition, Cambridge University Press, Cambridge, 2018), p. 317.

[23] Adoption of the Paris Agreement, FCCC/CP/2015/L.9/Rev.1.

[24] See further D. Bodansky, 'The Legal Character of the Paris Agreement' 25(2) *Review of European Comparative and International Environmental Law* 142 (2016).

[25] See further the discussion in A. Frank (ed.) 'Paris Climate Agreement: Success or Failure?' *Cosmos & Culture—NPR Blog* (published electronically), 12 January 2016.

[26] Paris Agreement, Article 2(1)(a).

[27] Paris Agreement, Article 2(1)(a).

span expires (though bear in mind that it has been noted above that the international community has failed to bring Doha properly into force at the time of writing). A reason that the coming into force date of the Paris Agreement is slightly vague rather than specific is that the Paris Agreement at Article 21 stipulates that the point at which the agreement will enter into force depends on how fast participating countries complete their approval processes.[28] As to developing and embedding further targets for emissions reductions that could follow on meaningfully from the Kyoto Protocol phase, the Paris Agreement contains nothing of the sort, leaving parties to the Agreement to pre-pare, communicate and maintain their own 'Nationally Determined Contributions' in a bottom-up manner that gives each of the states sig-nificant discretion.[29] In the author's view, while the Paris Agreement tends to be celebrated as a meaningful achievement in terms of interna-tional *diplomacy*, as it should be, it is perhaps arguable that in terms of international *law* it cannot be said to truly represent a notable strength-ening of the gradually developing international climate law regime. Easing the pressure that had been placed on securing further hard-fought binding top-down emissions reduction targets in favour of tip-ping the balance towards affording parties more extensive bottom-up wriggle room means that international climate law may continue to remain substantially devoid of a level of desirable legal force in this area.

It is also noteworthy that the USA signed the Paris Agreement, which was far from a foregone conclusion given the USA's often resistant stance in this area of governance on the international stage. However, the country signed the agreement prior to American President Donald Trump's tenure in office, who has since gone on to roll back USA pro-tections in the area of climate and energy considerably.[30] Other espe-cially notable signatories of the Paris Agreement include Japan, which had binding commitments under the first Kyoto Protocol period but that refused to adopt commitments for the second 2013–2020 period.

---

[28] Paris Agreement, Article 21(1): 'This Agreement shall enter into force on the thirtieth day after the date on which at least 55 Parties to the Convention accounting in total for at least an estimated 55% of the total global greenhouse gas emissions have deposited their instruments of ratification, acceptance, approval or accession'.

[29] A good account of this tilting of the scales towards bottom-up state-determined discre-tion is provided over pp. 321–323 in Sands, et al., supra, n. 22.

[30] Inaugurated 20 January 2017.

In addition to signalling that Japan was reluctant to contain its proportionally high level of international emissions contributions, Japan's withdrawal from the Kyoto Protocol was also a significant blow to international climate governance on a symbolic level, in that the breakthrough Kyoto Protocol agreement itself was adopted in Japan (at Kyoto) on 11 December 1997. A further particularly important signatory to the Paris Agreement includes the Russian Federation; like Japan, the Russians did have binding commitments under the first Kyoto Protocol period but had refused to adopt binding commitments for the second period.[31]

It is by now clear both that the international arena contributed significantly to the form and nature of the EU's climate law regime, and that the EU's own international presence and input contributed in turn to the development of international governance. As an EU Member State, the UK has for a long time been subject to these channels of influence from its largely nested position within the EU.[32] Subscription to the UNFCCC, the Kyoto Protocol and so on clearly embody 'direct' contributions to the UK climate and energy governance regime, insofar as the international experience has long had a formative influence on UK climate and energy governance through these channels. As such, it is notable that the CCA in some respects reflects the unfolding international governance agenda, having been consciously designed in a flexible way in order to absorb international requirements, e.g. through the CCA's flexible carbon budget regime, its capacity to harmonise with international trading schemes, its standardisation of targeted greenhouse gases with international definitional and reporting norms, etc.[33] Nonetheless, the CCA's subservience to international developments should not be overstated: it remains, at

[31] S. Schiele, *Evolution of International Environmental Regimes: The Case of Climate Change* (Cambridge University Press, Cambridge, 2014), p. 87. New Zealand is a further proportionally high-level emitter that supported the Paris Agreement after having agreed to targets under the first Kyoto Protocol period but rejecting them under the second period.

[32] Although, as noted, the UK initially participated in the UNFCCC from a largely nested position within the EU, the UK was also party to the agreement in its own right, meaning that Brexit—the process of the UK leaving the EU—has no capacity to sever the UK's status as a UNFCCC signatory. The same is true for the Paris Agreement; see further Lord Bourne of Aberystwyth, 'The Paris Agreement proves that the Transition to a Climate-Neutral and Climate-Resilient World is Happening', published speech transcript to the UN, delivered 22 April 2016, published 25 April 2016 (UK Government, 2016).

[33] On the key features of the CCA, see further Chapter 2.

heart, an internal UK innovation and at times has persisted in spite of international developments rather than because of them. In other words, the locked-in long-term nature of the framework has served to provide a robust steadying hand on the tiller of UK climate mitigation and adaptation processes at many points where the international arena was deadlocked over the course of negotiations, or where some progress was made but the outcomes were weak (e.g. where weak policy documents were agreed upon rather than robust emissions reduction targets).

2

In addition to the UK's 'broader' international experience, its narrower international experience within the internal world of the EU itself, which the UK joined in 1973, has also served to provide a very particular background setting for the creation and application of the CCA. The EU 'level' of governance is often described as 'supranational' rather than 'international', in acknowledgement of the EU's distinct and particular form and nature. Although the UK triggered Article 50 of the Treaty on European Union on Wednesday 29th of March 2017 and thereby set in motion the UK's exiting of the EU, known as 'Brexit',[34] aspects of the EU's internal formative influence on the UK while it was a conventional EU Member State should be acknowledged in any detailed consideration of the CCA's international context. Furthermore, it is also important to recognise that the CCA itself has exerted some notable degree of outward impact of its own on the broader international vista. The unfolding analysis below will develop the preceding international section by telescoping in to consider the narrower internal EU setting that provided a nuanced international governance context in its own right for the UK as it crafted and applied the CCA. The analysis will then acknowledge and address a sense of international influence that the CCA has itself generated amongst the wider international community.

While the EU was engaging externally with the international vista considered above, it was also working to develop its own internal climate and energy regime. As noted, progressive international agreement was

---

[34]On the UK's 2016 referendum to leave the EU, and the social, geographic and ideological dimensions underpinning the voting process, see M.J. Goodwin, O. Heath, 'The 2016 Referendum, Brexit and the Left Behind: An Aggregate-Level Analysis of the Result' 87(3) *The Political Quarterly* 323 (2016).

hampered by a protracted inability to reach wide-ranging consensus on ambitious emissions reduction targets and associated matters, and this occurred in spite of some significant pressure applied by the EU towards these ends internationally.[35] While the EU's call for more robust action was left to some extent frustrated on the international stage, the EU was at greater liberty to press ahead with a stronger decarbonisation regime within its own internal policy arena.[36] In particular, as the 2007 Bali Road Map was being cohered, and with EU diplomats continuing to push for reduction commitments at the international level, the EU was also pressing on internally by developing and rolling out a fairly sophisticated and relatively stringent climate and energy governance regime of its own. These forces have formed both a part of the CCA's initial context and governance setting, and interacted to some extent with its actual substance.

It was reasoned by the EU, and including by the UK, that if energy decarbonisation was to contribute to an EU Low Carbon Transition adequately, then a particular series of assumptions would need to underpin a fit-for-purpose EU decarbonisation regime. The headline assumptions included a transition away from finite sources of energy by embracing the opportunities and benefits afforded by renewable energy sources[37]; a suppression where possible of heavy greenhouse gas-emitting fossil fuels; a drive towards greater energy efficiency, including in particular a macro-focus on driving down general energy demand levels in addition to micro-policies such as applying targeted high standards of energy efficiency to new-build homes[38]; and stimulation of low carbon

---

[35] Good consideration of EU participation in the UNFCCC context is provided in S. Schunz, 'The EU in the United Nations Climate Change Regime', Chapter 10 in J. Wouters, H. Bruyninckx, S. Basu, S. Schunz (eds.) *The European Union and Multilateral Governance: Assessing EU Participation in United Nations Human Rights and Environmental Fora* (Palgrave Macmillan, Basingstoke, 2012).

[36] S. Oberthur, C.R. Kelly, 'EU Leadership in International Climate Policy: Achievements and Challenges' 43(3) *The International Spectator* 35 (2008).

[37] See further G. Boyle, *Renewable Energy: Power for a Sustainable Future* (Oxford University Press, Oxford, 2012).

[38] It is to be noted that the vital issue of energy efficiency in EU environmental and energy studies tends to be generally under-represented. Good coverage is provided in J. Rosenow, F. Kern, 'EU Energy Innovation Policy: The Curious Case of Energy Efficiency', Chapter 28 (pp. 501–518) in R. Leal-Arcas, J. Wouters (eds.) *Research Handbook on EU Energy Law and Policy* (Edward Elgar, Cheltenham, 2017).

technological development, especially Carbon Capture and Storage capabilities. Taken cumulatively, this broad EU-level rationale both echoed and reinforced governance drivers within the UK that were serving to galvanise the design of the CCA in the lead-up to its enactment.

The EU endeavoured to reflect these understandings in a specially tailored 2020 climate and energy programme, the *20-20-20 Programme*.[39] This framework has since been extended to 2030, in order to harmonise with longer-term EU energy intentions.[40] The *20-20-20 Programme* was published at a point in time in which the international community had been struggling to draw the 2007 Bali Road Map together, and in the period where the drive towards the creation of the CCA was building up an ineluctable head of steam inside the UK.[41] Here, the EU's European Council announced that by the year 2020 the EU was making a political commitment to: the reduction of EU greenhouse gas emissions by 20% below a 1990 measurement baseline; the improvement of energy efficiency by 20%; and the increase of renewables in the EU energy mix by 20%.[42] The commitment to attaining these targets by the year 2020 lent weight to the perceived importance of including that date as a core interim target within the UK's CCA.[43]

Perhaps indicative of the UK's ambition in this area, the UK's CCA was already firmly in place when EU legislation appeared in 2009 to translate the *20-20-20 Programme* intentions into binding EU legal obligations. In terms of the substance of the EU-level legislative framework, it was—and still is at the time of writing—built on an Emissions Trading Scheme Directive (as revised), requiring carbon emissions to be cut from regulated industry by 21% from a 2005 baseline level by 2020.[44] Areas outside of the scheme's remit (e.g. transport and housing) are caught by an Effort

---

[39] Often described as the *2020 Climate and Energy Package*, the *20-20-20 Targets/Goals*, or similar.

[40] See further: European Commission, *A Roadmap for Moving to a Competitive Low Carbon Economy in 2050* (COM/2011/112); European Commission, *Energy Roadmap 2050* (COM/2011/885); European Commission, *A Policy Framework for Climate and Energy in the Period from 2020 to 2030* COM (2014) 15 final/2.

[41] See further Chapter 1.

[42] Council of the European Union, 'Presidency Conclusions', Brussels European Council 8/9 March 2007 7224/1/07 CONCL 1.

[43] This became the date that the CCA's 'interim' milestone target has been pegged to, applied at CCA, s.5(1)(a).

[44] Council Directive 2009/29/EC [2009] OJ L140/63 ('ETS Directive').

Sharing Decision, which in the UK's case passed down a requirement to reduce UK emissions by 16% from 2005 baselines by 2020.[45] The law on renewables was asserted by the Renewables Directive, which included the obligation to achieve a total EU renewable energy share of 20% by 2020.[46] In applying this, the Directive passed varied target percentages to each EU Member State.[47] For the UK, the national renewables share was to be lifted from 1.3% (measured at a 2005 baseline) to 15% by 2020.[48] This intersected neatly with the CCA's 2020 target, helping the UK Act here, as elsewhere, to absorb elements of the EU-level requirements relatively seamlessly. Further, a Carbon Capture and Storage Directive was enacted in order to stimulate investment in and deployment of Carbon Capture and Storage technology,[49] and energy efficiency was treated under existing EU law[50] and then further reinforced three years later under a tailored Energy Efficiency Directive.[51]

The CCA's flexibility, strategic long-term design and relatively stringent standards, which were generally stronger than the EU average,[52] meant that the UK was well placed to absorb the *20-20-20 Programme* drivers and permit them to cohere with its extant national framework. It is clear that the CCA was at heart a UK-driven regime, but one that was partially impacted and actively shaped by the rich environmental governance setting provided by the broader EU prior to Brexit. Thus, although the UK regime is led by the national level, and locked in through national legislation and policy, its form and nature have to at least some extent been galvanised from 'above' by the international setting (supranationally by the EU, and internationally by the COP and associated

[45] Council Decision 406/2009/EC [2009] OJ L 140/136 ('Effort Sharing Decision'); 16% reflects the UK-specific target at Annex II to the Decision.

[46] Council Directive 2009/28/EC [2009] OJ L140/16 ('Renewables Directive').

[47] Renewables Directive, ibid., Annex I.

[48] Renewables Directive, ibid., Annex I.A.

[49] Council Directive 2009/231/EC [2009] OJ L140/114 ('CCS Directive').

[50] Council Directive 2002/91/EC [2002] OJ L1/65; Council Directive 2010/31/EU [2010] OJ L153/13; Council Directive 2006/32/EC [2006] OJ L114/64; Council Decision 1639/2006/EC [2006] OJ L 310/15; Council Directive 2009/125/EC [2009] OJ L285/10; Council Directive 2010/30/EC [2010] OJ 153/1.

[51] Council Directive 2012/27/EU [2012] OJ L315/1 ('Energy Efficiency Directive').

[52] See, e.g., B. Ward, 'How Will Brexit Affect Climate Change Policy?' *News & Commentaries* (published electronically, unpaginated; Grantham Research Institute on Climate Change and the Environment) 30 June 2016.

governance arenas). On first inspection of the CCA, these multilevel factors are not immediately apparent, insofar as the CCA has the superficial appearance of a self-contained UK-specific framework, but where one seeks to probe or assess the mitigation and adaptation processes created and applied by the regime in broader detail, these wider factors can be seen to exert a latent formative influence that can be drawn out.

In addition to recognising the impact of the international level on the CCA, it is also important to acknowledge that there is some degree of a two-way channel at play here: the impact of the international level on the UK framework has been unquestionably significant; however, the CCA itself has also received significant international attention in its own right, and, moreover, exerted some degree of international influence itself. Fankhauser, Averchenkova and Finnegan conducted a review of the CCA ten years after its enactment,[53] incorporating insights from interviews with a range of key (anonymous) stakeholders. The report finds that one of the CCA's prominent 'areas of success' is that the 'UK's international standing has grown':

> The CCA was one of the first comprehensive climate laws adopted globally and became the basis of a sustained international campaign on climate change by the Foreign Office. That engagement was one of the unexpected successes of the Act, helping the UK to play a leadership role in negotiating the Paris Agreement and inspiring other countries to take action.[54]

It is beyond the parameters of this book to undertake a comprehensive evaluation of the CCA's broader international influence, though these observations point the way towards interesting future research projects in the sphere of environmental studies. Nevertheless, it is possible in the context of the present study to offer something of a flavour of the CCA's international impact. Here, it is to be remembered that the CCA was the first national climate framework of its kind in the world, and as such it is unsurprising that the UK model has attracted a good deal of attention from parties and actors in the international community working in this area.

---

[53] S. Fankhauser, A. Averchenkova, J. Finnegan, *10 Years of the UK Climate Change Act* (Grantham Research Institute and London School of Economics, 2018).

[54] Fankhauser, Averchenkova, Finnegan, ibid., p. 3.

This international attention has gone beyond general interest or observation,[55] to include strategic policy interest[56] and practical political-legal development. In terms of this latter point, it is the case that a number of countries now employ climate regimes containing foundational elements that have been influenced directly by the UK approach. Thus, as the UK's CCC has emphasised in a commentary piece entitled *A Global Deal of National Climate Change Laws?*, 'both Denmark and Finland have adopted new climate change laws. Both countries [*sic.*] Climate Change Acts are modelled closely on the UK's'.[57] The CCA has played a similarly influential role in the context of climate legislation established in Sweden. The Swedish Minister for Financial Markets and Consumer Affairs has described the CCA as acting as 'a guiding star' when Sweden was considering how best to fashion its own national climate legislation.[58] The CCA has also featured at the forefront of a push for an equivalent type of Act in New Zealand, called the Zero Carbon Act. Lang has noted as follows:

> Modelled on the UK's 2008 Climate Change Act – which has an 80% emissions-reduction target to 2050 – the Zero Carbon Act will commit New Zealand to a 100% emissions-reduction target to 2050, ensuring they are carbon neutral by 2050 – or thereabouts.[59]

[55] General interest in the CCA is reflected in media and associated coverage that the CCA has received outside the UK.

[56] An example of 'Strategic Policy Interest' includes the 'Outcome Document' produced at the event 'International Environmental Laws and Climate Change: Is there Global Climate Justice', hosted 17–20 July 2017 by the Center for United Nations Constitutional Research (CUNCR). This outcome report, which forms a basis for UN-level lobbying work conducted by CUNCR, states that 'Domestically, in passing the first rigorous climate change legislation in 2008, the UK has provided the world with a crucial opportunity to learn from its successes and failures—not least given its complex constitutional environment. The 2008 Climate Change Act is a useful model for how other States can implement legally binding targets to combat climate change'. See CUNCR, *Is there Global Climate Justice?—Outcome Document* (CUNCR, 2017).

[57] CCC, 'A Global Deal of National Climate Change Laws?' published online at the CCC's database, 23 June 2014, https://www.theccc.org.uk/2014/06/23/a-global-deal-of-national-climate-change-laws/.

[58] Minister Per Boland, in conversation with the author over the course of research for this book, on the date of Friday 20 October 2017.

[59] J. Lang, 'Zero Time: NZ's Zero Carbon Act' *E-nvironmentalist* (published electronically, unpaginated) 18 September 2017.

Perhaps the most profound identifiable influence in the first instance occurred in relation to Mexico, whose officials considered and studied the CCA in detail. Ultimately, it influenced Mexico as a legislative template, and was drawn on to help the country develop its own pioneering General Law on Climate Change.[60] The International Development Law Association has outlined that:

> Mexico passed the General Law on Climate Change on April 19, 2012, establishing a new leading global legal best practice to address climate change. Mexico has become the second country, after the United Kingdom, to set out a regulatory framework that comprehensively addresses climate change through a committed multi-sectoral and multi-stakeholder approach.[61]

It is also notable that in passing this important framework, Mexico established 'the first comprehensive national law to address climate change enacted *in the developing world* and the second one at the global level… after the Climate Change Act, passed in the United Kingdom in 2008[.]'[62] The UK's long-term targets and shorter reporting periods also exerted some significant influence at the international level as the Paris Agreement was being constructed.[63] In addition to influence of this kind in the United Nations COP environment, and bilateral influence, as in the case of Mexico, further influence has also arisen through British Commonwealth channels. This has been the case, for example, in Papua New Guinea, where the nation's Climate Change (Management) Act 2015 displays some echoes of the CCA.[64]

---

[60] See, e.g., M. Nachmany, S. Fankhauser, et al. *The 2015 Global Climate Legislation Study: A Review of Climate Change Legislation in 99 Countries, Summary for Policy-Makers* (Grantham Research Institute, GLOBE, IPU, 2015), p. 15.

[61] IDLO, *The New General Law on Climate Change in Mexico: Leading National Action to Transition to a Green Economy* (IDLO, 2012), p. 1.

[62] *The New General Law on Climate Change in Mexico*, ibid., p. 4 (emphasis added).

[63] Stephen Minas (expert in international climate affairs), in conversation with the author over the course of research for this book, on the date of Wednesday 11 October 2017; see also S. Fankhauser, A. Averchenkova, Finnegan, supra, n. 53.

[64] Climate Change (Management) Act 2015. See also Papua New Guinea Office of Climate Change and Development, *Papua New Guinea's Commitment to Act on Climate Change* (Papua New Guinea Government, 2010). The author is grateful to Stephen Minas for highlighting Papua New Guinea's circumstances over the course of research for this book.

While further targeted research that ranges substantially beyond the scope of this book would be required in order to more fully assess the direct influence that the CCA has exerted on law and policy regimes around the world, it is clear nevertheless that the CCA has exerted some significant degree of influence in the international arena. Associated with this, and as a point of general principle, it is also clear that the CCA holds lessons in the area of climate governance and environmental studies that the international community would be well advised to take note of. Thus, the extent to which the CCA might serve as an inspiration for the creation of robust climate frameworks elsewhere in the world has relevance. Beyond this, the CCA's capacity to serve as a direct model or template for other states or international-level actors seeking to create their own environmental frameworks should also be recognised. This recognition suggests that both the international utility and applicability of the Act, in addition to its weaknesses and pitfalls, such as they may be, should be borne in mind as an important point of reference by the wider international community as it moves forward in combatting the climate crisis.

3

The preceding sections of this chapter have considered interaction between the UK's national-level CCA and the international level, and the analysis is now well poised to address the relationship between the national CCA and the UK's subnational level of governance. Nation states are highly complex constructs, and the UK's subnational governance arena is highly complex in its own right. The present consideration will hone in on the UK's devolved tier of governance, where, beneath the level of UK Parliament and UK Government, the UK's devolved territories of Northern Ireland, Scotland and Wales each operate their own devolved governments and parliaments. England also amounts to a major substate UK region; however, it does not operate a devolved Parliament or government, and so the focus here remains on the devolved territories and their substate governmental apparatuses. These intricate devolved governance circumstances raise special nuanced questions, open opportunities and create challenges for the CCA's 'national' climate and energy mitigation and adaptation regime.

As just noted, special devolved parliaments and governments operate below the UK's national level in Northern Ireland, Scotland and

Wales. These contemporary devolved arrangements were established in 1998.[65] The arrangements mean that UK Parliament legislates on national matters, and that the devolved parliaments can create legislation in areas of competence that national Parliament has devolved to them. Thus, below the national level, a devolved Northern Irish Parliament operates at Stormont in Belfast, called the Northern Ireland Assembly. Scotland operates the Scottish Parliament at Holyrood, in Edinburgh, and Wales' equivalent Parliament is called the National Assembly for Wales, operating in Cardiff.[66] Sometimes, national Acts created by UK Parliament will cut into devolved competences, and so, under a legislative consent convention, frequently described as the 'Sewel Convention', a provision of an Act of UK Parliament that encroaches upon devolved powers will extend to the devolved jurisdiction(s) in question only if the devolved Parliament(s) has passed a motion consenting to it. 'Energy' is chiefly a national-level competence retained by UK Parliament, while 'environment', which also catches aspects of climate governance, not least adaptation, tends to be significantly devolved[67]; each of the devolved jurisdictions has consented to the CCA and is subject to its terms.

Coming first to Wales, the Welsh administration is subject to the CCA (as is the case with Northern Ireland and Scotland), and it remains the primary climate mitigation and adaptation legislative framework operating on Wales. As a quirk of technical UK constitutional law, it is the case that Northern Ireland and Scotland are individual jurisdictions within the UK, whereas there is technically no individual jurisdiction of Wales in spite of

---

[65] The underpinning legislation was enacted in 1998, and the arrangements had become properly operational by 1999. However, unlike Northern Ireland and Scotland, in the case of Wales a fully developed executive Welsh government that was separate from the Welsh legislature was not created in law until 2006.

[66] UK devolution was established by a range of Acts created by UK Parliament, most particularly the Northern Ireland Act 1998, the Scotland Act 1998 and the Government of Wales Act 1998. The devolved settlements have changed and evolved over time, with the changes being brought in by various legal amendments and additional Acts of UK Parliament. The Welsh arrangements were substantially adjusted in particular under the terms of the Government of Wales Act 2006 and the Wales Act 2017.

[67] See further T.L. Muinzer, G. Ellis, 'Subnational Governance for the Low Carbon Energy Transition: Mapping the UK's "Energy Constitution"' 35(7) *Environment and Planning C* 1176 (2017).

its devolved Assembly.[68] Instead, due to a degree of technical legal conformity between Wales and England, Wales forms part of the single jurisdiction of 'England and Wales'.[69] Therefore, and given that the present work is written in the broader tradition of environmental studies, rather than as a narrow technical law text, it should be noted that the term 'jurisdiction' is sometimes used here as a rhetorical device to describe the UK's subnational regions in general; however, lawyers and legal scholars consulting this text should bear in mind that within the discrete confines of their technical discipline Wales does not constitute a free-standing jurisdiction in the way that Northern Ireland and Scotland do. The foundations of Wales' current devolved arrangements were laid by the Government of Wales Act 1998 (hereafter 'GoWA 1998'), which established the present National Assembly for Wales. As part of this process, the Act transferred powers that had been vested in UK Government's Secretary of State for Wales to the newly created Assembly.[70] This governance architecture was substantially adjusted under the terms of the Government of Wales Act 2006 (hereafter 'GoWA 2006'), and large swathes of the GoWA 1998 were repealed in the process[71]; in particular, the GoWA 2006 established an executive Welsh Assembly Government that was separate from the Welsh legislature, that is to say, separate from the National Assembly for Wales.[72]

Additional adjustments following in the wake of a Welsh Referendum held in March 2011 enabled the Welsh authorities to further extend the reach of their legislative capacity, empowering Wales to create law in

---

[68] For a discussion of the single jurisdiction circumstance in the light of certain comments on the subject from the First Minister of Wales, see M. Navarro, 'A Substantial Body of Different Welsh Law: A Consideration of Welsh Subordinate Legislation' 33(2) *Statute Law Review* 163 (2012).

[69] An accessible discussion of issues raised by these 'England and Wales' jurisdictional circumstances can be read in R. Owen, 'Should Wales Separate from England's Legal System', *The Conversation* (published electronically, unpaginated) 12 April 2016.

[70] See further: D. Lambert, 'The Government of Wales Act—An Act for Laws to Be Ministered in Wales in Like Form as It Is in This Realm?' 30 *Cambrian Law Review* 60 (1999); M. Laffin, A. Thomas, 'Designing the National Assembly for Wales' 53(3) *Parliamentary Affairs* 557 (2000).

[71] R. Owen, 'Government of Wales Act 2006' 42(1) *The Law Teacher* 103 (2008).

[72] A. Trench, 'The Government of Wales Act 2006: The Next Steps on Devolution for Wales' *Public Law* 687 (2006).

twenty specific areas of devolved competence.[73] In spite of the importance of these developments, and a tradition of innovative thinking and action in certain areas of governance, Wales' actual devolved climate and energy powers have been limited.[74] The GoWA 2006 and associated law broke the Welsh devolved powers into twenty devolved 'subjects', which have included some energy and broader climate capacities, but all in all they have not been very far-reaching.[75] In terms of Wales' overall policy approach within the architecture of the CCA, the emissions reduction drive is supported by a *Climate Change Strategy for Wales*, which includes the intention to '[r]educe greenhouse gas emissions by 3% per year from 2011 in areas of devolved competence, against a baseline of average emissions between 2006–10'.[76] The energy sector is specially targeted as a means of supporting this process.[77]

In spite of its relatively narrow range of powers, the National Assembly for Wales has succeeded in employing subnational law as a means of taking some innovative action, for example by passing the Well-Being of Future Generations (Wales) Act 2015, which has exerted influence on Welsh climate- and energy-related decisions and strategies.[78] Furthermore, Wales has passed the Environment (Wales) Act 2016, with Part 2 to this Act containing substantial climate change provisions.[79] These provisions place the Welsh Ministers in the substate government

---

[73] See GoWA 2006, Schedule 7, Subjects 1–20. Further significant amendments have since been made to the Welsh legislation by the Wales Act 2017.

[74] See further the analysis of the Welsh powers across T.L. Muinzer and G. Ellis, supra, n. 67. See also R. Cowell, 'Decentralising Energy Governance? Wales, Devolution and the Politics of Energy Infrastructure Decision-Making' 35(7) *Environment and Planning C: Politics and Space* 1242 (2017).

[75] GoWA 2006, Schedule 7 Part 1. 'Fields' wherein the Welsh Assembly has been able to make law were established at GoWA 2006, Schedule 5, and GoWA 2006, s.108 limited these to a range of subjects stated at GoWA 2006, Schedule 7. Additional limits described as 'exceptions' within Schedule 7 were placed on certain subsidiary aspects of the subjects. See also the Wales Act 2017, where further significant adjustments to the governance structure of the Welsh framework have been introduced.

[76] Welsh Assembly Government, *Climate Change Strategy for Wales* (WAG, 2010), p. 34.

[77] See Welsh Assembly Government, *Energy Wales: A Low Carbon Transition* (WAG, 2012).

[78] The legislation affords special recognition to redressing problems posed by climate change, see, e.g., Well-Being of Future Generations (Wales) Act 2015, s.4, Table 1, Goals one and two. Elements of the CCA are also explicitly recognised at s.11(2)(b) and s.38(3)(a).

[79] Environment (Wales) Act 2016, Part 2, 'Climate Change'.

under a duty to issue regulations that establish interim emissions targets for 2020, 2030 and 2040 in the lead-up to the CCA's 2050 80% emissions reduction marker.[80] The Ministers are also placed under a duty to issue regulations that establish 5-year carbon budgets for Wales, which, again, reflect substate absorption of the equivalent CCA requirements, and are to be pegged to the CCA 2050 milestone target trajectory.[81] These powers, and other reporting, advisory and procedural duties in the vein of the CCA that interact with them appear across Part 2 to the Act, which is introduced by a section entitled 'Purpose of this part'.[82] Here, it is stated that the purpose of Part 2 is 'to require the Welsh Ministers to meet targets for reducing emissions of greenhouse gases from Wales'.[83] As such, the Act can be seen to be opening up a substate channel in legislation that permits the Welsh to work in a targeted and sophisticated way towards the CCA outcomes.[84]

As noted above, the CCA is also the primary climate and energy mitigation and adaptation legislation applying in Northern Ireland. Northern Ireland's Devolved Administration has adopted a soft policy approach to climate governance within the umbrella of the CCA requirements, where it has framed its goals around an expressed intention to reduce greenhouse gas emissions by 35% on 1990 levels by 2025.[85] This commitment has been actively driven by the CCA, which, as has been seen, commits

---

[80] Environment (Wales) Act 2016, s.30, s.32.

[81] Environment (Wales) Act 2016, s.31, s.32.

[82] Environment (Wales) Act 2016, s.28.

[83] Environment (Wales) Act 2016, s.28.

[84] The Environment (Wales) Act 2016 also devotes a major section to the absorption of CCA-driven carrier bags charges (discussed in the preceding chapter); see Environment (Wales) Act 2016 Part 3 ('Charges for Carrier Bags') and Schedule 2 Part 2.

[85] See Northern Ireland Executive, *Programme for Government 2011–15* (NIE, 2011). At the time of writing, the Northern Irish government has collapsed, and this has delayed the finalisation and implementation of a subsequent *Programme for Government*. Ironically, the government's collapse was chiefly precipitated by climate and energy-specific issues, relating to the Democratic Unionist Party's alleged mismanagement of financial incentives for renewable heat (the Renewable Heat Incentive); see further T.L. Muinzer, 'Incendiary Developments: Northern Ireland's Renewable Heat Incentive, and the Collapse of the Devolved Government', (99) March/April *UKELA E-Law* 18 (2017). The draft version of the next *Programme for Government* gives no indication that a more robust approach will be taken to climate and energy mitigation and adaptation in the event that the devolved government is restored.

the UK as a whole to emissions reductions of 34% on 1990 levels by 2020.[86] Clearly, in working towards 2025 rather than 2020, Northern Ireland's target falls some way short of the national framework's aggregated normative standard.

The reasons for Northern Ireland's slack approach are not straight-forward. A general progressive will to embrace the CCA's long-term national targets in a manner that might permit Northern Ireland to equal or indeed exceed normative UK standards has not been broadly forthcoming across the Northern Irish political class[87]; there can be no doubt that Northern Ireland's administration has been the least progressive primary governance actor in the sphere of climate and energy mitigation and adaptation in the UK.[88] These circumstances are aggravated by additional factors, particularly by the way in which Northern Irish decarbonisation is impacted by the jurisdiction's specific nuanced socio-economic and geographic circumstances, which include amongst many complex variables an unusually high proportion of emissions sourced from the agricultural sector.

In relative terms, agriculture is a proportionally larger socio-economic sector in Northern Ireland than it is elsewhere in the UK. In its *Agriculture in the United Kingdom 2016* statistical publication, UK Government's Department for Environment, Food and Rural Affairs pointed out that Northern Ireland's agricultural sector employs a significantly larger relative proportion of the Northern Ireland workforce in comparison with relative employment proportions in the UK's other substate territories (and including England).[89] These sorts of circumstances trace through directly into emissions challenges, with the CCC highlighting that 'Northern Ireland has relatively high shares of

---

[86] CCA, s.5(1)(a).

[87] See, e.g., G. Ellis, R. Cowell, F. Sherry-Brennan, P. Strachan, D. Toke, *Delivering Renewable Energy Under Devolution: Initial Findings Summary Report* (DREUD, 2013), p. 6.

[88] See Muinzer's commentary on the tension between the fairly extensive capacities to take action in the area of climate and energy governance enjoyed by Northern Ireland under the devolution arrangements, and the fundamental lack of progressive governance action that the devolved administration has exhibited in practice: T.L. Muinzer, 'Warming Up: Northern Ireland's Developing Response to Climate Change in the Context of UK Devolution', (96) (September/October) *UKELA E-Law* 19 (2016); Muinzer, 'Incendiary Developments', supra, n. 85.

[89] DEFRA, *Agriculture in the United Kingdom 2016* (HM Government, 2017), p. 21.

emissions from agriculture and land use, land use change and forestry... and relatively high per capita emissions in agriculture... compared to the UK as a whole'.[90] Agriculture is amongst the most difficult sectors to significantly decarbonise and may indeed be the most difficult given the current state of knowledge and available practical decarbonisation mechanisms. Further, Northern Ireland is also undergoing a post-conflict transition, where it has recently moved out of a long period of violent sociopolitical conflict known as 'the Troubles', an era that has had a detrimental effect on stable and coherent governance in the jurisdiction.[91]

At any rate, in general terms there is a strong sense of meaningful contrast between Northern Ireland's 'soft' policy approach to decarbonisation, as embodied by its '35% by 2025' policy reduction commitment noted above, in comparison with the 'hard' national objectives embodied in law by the CCA targets and associated duties. Numerous calls have arisen for the Northern Ireland Assembly to legislate a bespoke subnational Climate Change Act that could reinforce the national CCA in a meaningful way within Northern Ireland, including calls from NGOs such as *Friends of the Earth (Northern Ireland)*, expert individuals and bodies,[92] favourable recommendations to that end from the CCC itself[93] and support for the creation of binding statutory targets from successive Northern Ireland Environment Ministers.[94] A public consultation on a Northern Ireland Climate Bill has taken place, which resulted in

[90] CCC, *The Appropriateness of a Northern Ireland Climate Change Act—December 2015 Update* (CCC, 2015).

[91] See D. McKittrick, D. McVea, *Making Sense of the Troubles* (New Amsterdam Books, USA, 2002).

[92] See, for example, the support offered by the Royal Town Planning Institute Northern Ireland (RTPINI) in *Pre-consultation Seeking Views on the Need for a Northern Ireland Climate Change Bill: A Response by the Royal Town Planning Institute Northern Ireland* (RTPINI, 2013).

[93] CCC, *The Appropriateness of a Northern Ireland Climate Change Act* (CCC, 2011), pp. 18–19.

[94] Alex Atwood MLA (Minister for Environment, May 2011–July 2013); see, e.g., M. McKimm, 'Northern Ireland "Should Do More over Carbon Emissions"', *BBC News*, 8 November 2011. Atwood's successor, Mark H. Durkan MLA (Minister for Environment, July 2013–March 2016), was also sympathetic to this view, describing Northern Ireland's inability to reach political consensus in order to legislate in this area as an 'embarrassment' to Northern Ireland's devolved Parliament; see "Environment Minister Rules Out Separate Climate Change Laws', *Belfast Telegraph* (UK Newspaper), 5 December 2016.

very strong support overall from respondents for the introduction of a Climate Bill,[95] however sufficient support across key political actors within the Devolved Administration itself could not be mustered to the point that the necessary legislation could be taken forward.

In 'Northern Ireland's Consent to the Climate Change Act 2008: Symbol or Illusion?' Turner has argued that Northern Ireland's consent to the terms and application of the CCA looked on the face of it like it represented a long-term commitment from diverse Northern Irish political parties to share power in Northern Ireland in the interests of present and future generations, and as such it also appeared to indicate a broader commitment to participate in positive, progressive devolved government within the jurisdiction.[96] She goes on to suggest, however, that in reality this consent was semi-procedural insofar as it was effectively necessitated by Northern Ireland's regional subjugation to national authority; in this respect, the Devolved Administration's active, genuine consent to the CCA was therefore potentially illusory.[97] In other words, instead of using its relatively weighty devolved powers to take a degree of *active ownership* over Northern Irish decarbonisation, including in the area of energy, where Northern Ireland enjoys unusually substantial devolved controls, the Devolved Administration has established instead a culture of 'soft' and relatively inert governance in this sphere, which has arisen in no small part through the largely *passive acceptance* of the CCA's mandatory requirements.

Scotland's Devolved Administration and Parliament have reacted differently to the CCA, by transposing its requirements into a highly sophisticated, legally binding substate framework of its own. Indeed, in doing so the Scottish Parliament has actually chosen to ramp up certain key aspects of the CCA duties within Scotland. As with Northern Ireland and Wales, the CCA legally binds on Scotland, as one part of the overall

---

[95] See DOE, *Synopsis of Responses to the Department's Pre-consultation Seeking Views on the Need for a Northern Ireland Climate Change Bill*, (DOE, 2013). Here, it is recorded that 64% of respondents favoured the creation of a Climate Change Bill for Northern Ireland, 24% were against the Bill, and 12% did not express a preference (p. 7).

[96] S. Turner, 'Northern Ireland's Consent to the Climate Change Act 2008: Symbol or Illusion?' 25(1) *Journal of Environmental Law* 63 (2013), p. 63. See also S. Turner, 'Committing to Effective Climate Governance in Northern Ireland: A Defining Test of Devolution' 25(2) *Journal of Environmental Law* 203 (2013).

[97] Turner, ibid., 'Northern Ireland's Consent to the Climate Change Act 2008', p. 63.

UK, obligating it amongst other things to contribute to the national reduction in greenhouse gas emissions of 80% below 1990 baselines by 2050 and to the associated 34% interim reduction target for 2020.[98] As noted, and notwithstanding its subservience to this framework, the Scottish Parliament has taken a pro-active position by legislating in this area in its own right: this does not mean that Scotland is no longer subject to the CCA, but means, rather, that while remaining subject to the national framework Scotland has used its devolved authority to adjust its own intentions and goals within the acceptable parameters of that framework. Scotland has achieved this through the creation of the Climate Change (Scotland) Act 2009 (hereafter 'CC(S)A'). This important subnational Act is partially modelled on the CCA, and it is sophisticated and significant enough to deserve targeted exploration and analysis in its own right. This targeted analysis is undertaken below, under the next numbered heading.

In general terms, Scotland has adopted a broadly progressive approach to climate and energy mitigation and adaptation, and the Scottish Government has indicated that the Scots are endeavouring to lead the way in the emissions reduction drive within the UK.[99] As explored in the next section, in the run-up to 2020 Scotland has adopted and is working towards a more stringent interim target than that required by the CCA under the terms of Scotland's key headline legislation (the CC(S)A), thus signalling in no small part that the Scottish way of 'doing climate and energy' can be a way that exceeds normative expectations.[100] In spite of Scotland's proportionally low population (some 5.4 million[101] out of a UK total of around 65.6 million people[102]), over a quarter of UK renewable electricity generation

[98] It is also noteworthy that it is possibly the case under constitutional law that national decarbonisation targets cannot be set without consultation with the Scottish Ministers having first taken place; see the Scotland Act 1998, s.106(4).

[99] See, e.g., Scottish Government (press release), 'Leading the Way on Climate Change', 10 November 2017.

[100] See, for example, Turner's discussion of the CCA, who considers Scotland to be an exemplary leader (in contrast to a comparatively recalcitrant Northern Ireland); Turner, supra, n. 96, 'Northern Ireland's Consent to the Climate Change Act 2008', p. 65.

[101] National Records of Scotland, *Mid-2016 Population Estimates Scotland* (Scottish Government, 2017).

[102] Office for National Statistics, *Overview of the UK Population: July 2017* (ONS, 2017), p. 2.

already originates in Scotland due to the jurisdiction's extensive renewables roll-out.[103] It is also possible to view the CC(S)A as signalling that the Scottish approach to climate governance amounts to a particularly equitable way of doing things, in the sense that environmental sustainability and anthropogenic climate change can be perceived as global challenges that need to be tackled in the interest of the common good, with Scotland leading the way in doing so within the UK. These sorts of interpretations are borne out in the Scottish Government's political rhetoric, which frequently celebrates Scotland's place at the international climate and energy vanguard.[104] However, one should perhaps not be too idealistic about these achievements, insofar as Scottish renewables have been heavily incentivised by UK Government from the public purse. These funds have acted as a means of stimulating renewables investment, meaning that in ramping up renewables within its jurisdiction the Scottish administration has been able to draw a high proportion of subsidy monies into its regional economy. Whatever one's view of the primary reasons driving Scotland's Low Carbon Transition, it remains the case that since the time of the CCA's creation the jurisdiction has forged ahead, coming quickly to embody one of the world's outstanding examples of substate governance.[105]

The following section develops the foregoing consideration of the UK's substate arena by providing the first targeted critique available in the environmental studies literature of Scotland's understudied but important CC(S)A. As necessitated by this book's subject matter, Scotland's CC(S)A is approached within the broader context of the UK's CCA. The section begins by providing more detail on the Scottish context in which the CC(S)A emerged, and then proceeds to explicate and analyse the 2009 framework itself.

---

[103] Scottish Government, 'Energy in Scotland: Get the Facts' (hosted at the Scottish Government's online database), http://www.gov.scot/Topics/Business-Industry/Energy/Facts (Based on readings for 2015).

[104] See, for example, the rhetoric underpinning the Scottish Government's 'Greener Scotland' website, http://www.greenerscotland.org.

[105] See further M. Peeters, M. Stallworthy, J. de Cendra de Larragán (eds.) *Climate Law in EU Member States: Towards National Legislation for Climate Protection* (Edward Elgar, 2012), in particular Chapter 7 (examining Scotland).

3

As noted above, given both its general subnational importance and its innovative legal-political design and characteristics, the CC(S)A deserves significant attention. Critiquing the framework is not, however, a particularly straight-forward task, insofar as the CC(S)A exhibits a level of complexity that is not entirely dissimilar in scope to that of the CCA. It has also received very little analytical and scholarly attention. In terms of the specific relationship between the CCA and the CC(S)A, it has been stressed above that the existence of the CC(S)A does not mean that Scotland is no longer subject to the CCA; it means, rather, that while remaining subject to the national framework the Scottish Parliament and Scottish Government have used their capacities for action within the UK's devolved constitutional setting to adjust Scotland's intentions and goals within permissible parameters. It is to be remembered that Scotland enjoys a substantial range of devolved powers, where it is at liberty to act and innovate.

Scotland may not, perhaps, be considered to be an especially geographically expansive subnational region when viewed in a global context, but within the UK itself its geographic range is such that it is the second largest substate jurisdiction in terms of land mass, after England. It also contains the second largest substate population, of around 5.4 million people.[106] As such, it is unsurprising that Scotland is a significant source of greenhouse gas emissions, e.g. producing approximately 48.1 $MtCO_2e$[107] over 2015.[108] Gauged at 1990 levels, the jurisdiction's emissions have been on a more pronounced downward trend than those of the UK generally, with the CCC reporting to Parliament in 2017 that 'Scotland is performing well compared with other countries in the UK and the UK as a whole. Scotland's

[106] As discussed in the preceding section of this Chapter.

[107] $MtCO_2e$ is a technical measurement meaning 'million tonnes carbon dioxide equivalent'. The $MtCO_2e$ reading quantifies the gases defined as 'greenhouse gases' under the Climate Change Acts.

[108] See Scottish Government, *Scottish Greenhouse Gas Emissions 2015* (Scottish Government, 2015), p. 2.

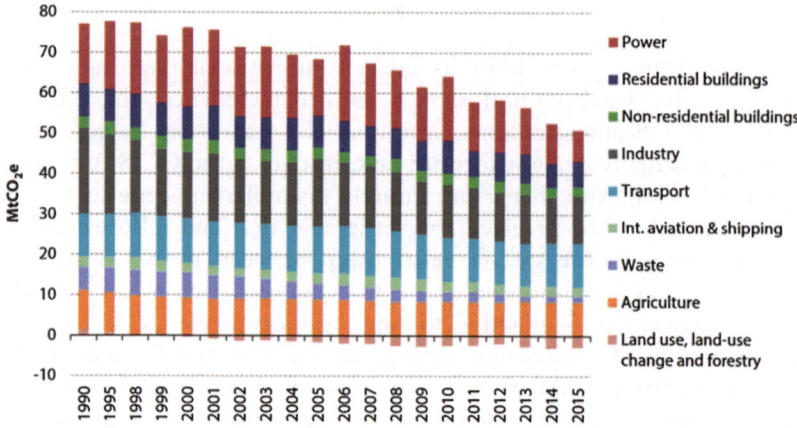

Fig. 3.1 *Source* UK Committee on Climate Change (2017): Cited in the Ecologic Institute Report '"Paris compatible" governance: Long-term policy frameworks to drive transformational change' (Ecologic Institute, 2017, p. 68)

actual emissions… in 2015 were 38% below 1990 emissions, compared with a reduction of 35% for the UK as a whole'.[109] The CC(S)A has played an important part in this process as a primary transition driver (Fig. 3.1).

That said, and as with the CCA itself, there are so many variables at play in the sphere of decarbonisation that one must be careful not to over-attribute pronounced emissions reduction successes to the CC(S)A. For example, while the CC(S)A has certainly had an important effect, it did not go 'live' in law until 2009, and as such it has not been operative over the full term of Scotland's post-1990 emissions reduction experience (and thus, it is certainly insufficient to fully explain it). In practice, one primary driver that has served to reduce Scottish emissions significantly since the 1990s has been a gradual policy decision to move towards the closure of Scottish coal-fired power plants. The last Scottish coal plant to close was Longannet power station, which had been the second largest power station in the UK until its closure in March 2016.

[109] CCC, *Reducing Emissions in Scotland: 2017 Progress Report to Parliament* (CCC, 2017), p. 7. The reason that 2015 readings are being used for reporting in 2017 is that emissions reporting is subject to a substantial time-lag due to the breadth and complexity of the data that must be gathered and processed.

The CC(S)A was introduced as a Bill to the Scottish Parliament in December 2008 by John Swinney MSP. It was amended over January–June of 2009, was passed by the Scottish Parliament on 24 June 2009 and received Royal Assent in August 2009. As with the creation of Acts in general in the UK, the symbolic act of Royal Assent meant that the Act had now been created into law. Amid a broad range of policies coloured with the hue of Scottish nationalism, the governing Scottish National Party (hereafter 'SNP') had been running on a 'greener Scotland' pro-environmental sustainability programme and were elected to power in Scotland in 2007. In addition to contributing to some significant extent to the Low Carbon Transition in Scotland in a general sense, the SNP's broader nationalist agenda has also served to galvanise a palpable will to *go beyond* England (and the rest of the UK) in this sphere, with Scotland often being cast as the most progressive of the decarbonising substate jurisdictions.[110] This politicised construction is not without some palpable foundation, as demonstrated by various practical indicators. In particular, such indicators include the type of measurement highlighted above in relation to quantifiable greenhouse gases, where it can be seen objectively that Scotland's emissions have decreased to a greater proportional extent than England's and the rest of the UK's.[111]

Further, additional 'legalistic' sorts of indicators can be identified in the law itself. The greatest 'headline' legislative indicator of Scotland's endeavours to 'go further' in the present context is the robust 42% interim reduction target for 2020 that has been embedded in the CC(S)A (see below).[112] This contrasts with the less far-reaching 34% 2020 target established under the CCA (which had initially been set at 26%[113]). Indeed, Scotland has gone so far in securing its 42% emissions reductions target well in advance of 2020 that this has prompted a

---

[110]See, e.g., H.V. Campbell, 'A Rising Tide: Wave Energy in the United States and Scotland' 2(2) *Sea Grant Law and Policy Journal* 29 (Winter 2009/2010). Jackson and Lynch have emphasised that the Scottish legislation imposes some of the most demanding legal requirements found not just in the UK, but anywhere in the world, see T. Jackson, W. Lynch, 'Public Sector Responses to Climate Change: Evaluating the Role of Scottish Local Government in Implementing the Climate Change (Scotland) Act 2009' 8/9 *Commonwealth Journal of Local Governance* 112 (2011), p. 112.

[111]Supra, n. 109, and text to note.

[112]CC(S)A, s.2(1).

[113]See further Chapter 1.

review of the legislation, in which (at the time of writing) the Scottish Government has issued a Climate Change (Emissions Reduction Targets) (Scotland) Bill for consultation. This may result in an increase of the 2020 target to 56% and an upward adjustment of the 2050 target to at least 90%.[114]

The SNP also insightfully connected CC(S)A developments to a broader 'green economic growth' agenda, where it was largely assumed that progressive decarbonisation and economic development could work in a mutually beneficial way. These circumstances can be contrasted with the Northern Irish experience, where economic development has been strongly acknowledged but climate-specific sustainability concerns have tended to be placed worryingly low on the agenda.[115] These dimensions are reflected in the SNP's *Programme for Government*, which connects the Scottish Low Carbon Transition with Scottish economic growth, with the First Minister of Scotland stressing in the 2017–2018 *Programme's* Foreword that 'We have a moral responsibility to tackle climate change and an economic responsibility to prepare Scotland for the new, low carbon world'.[116]

Coming now to the main elements of the CC(S)A itself, one major feature of the framework is that it lays down a legally binding target of a 42% reduction of greenhouse gas emissions for 2020 based on 1990 levels, and it also applies an equivalent long-term target of 80% below 1990 levels for 2050. As noted above, Scotland has secured its 2020 target at the time of writing, and the Scottish Government has signalled its intention to introduce amendments that will adjust the framework's target trajectory towards higher percentage reductions. At present, however, the 2020 and 2050 reductions remain as first stated in the law (42% and 80%, respectively, on 1990 levels). The '2050 target' is stated at Section 1, with the 'interim target'—namely the 2020 target—stated at Section 2.[117] Scotland's interim target is expressed in language that is

---

[114] See further Scottish Government, *Proposals for a New Climate Change Bill: Strategic Environmental Assessment Environmental Report* (Scottish Government, 2017), and the Climate Change (Emissions Reduction Targets) (Scotland) Bill.

[115] See further T.L. Muinzer, supra, n. 88.

[116] 'Foreword by the First Minister', in Scottish Government, *A Nation with Ambition: The Government's Programme for Scotland 2017–18* (Scottish Government, 2017), p. 3.

[117] CC(S)A, s.1(1), s.2(1).

slightly more clear and direct than the equivalent provision in the CCA, as follows:

> The Scottish Ministers must ensure that the net Scottish emissions account for the year 2020 is at least 42% lower than the baseline.[118]

These elements, as with much of the CC(S)A in general, are modelled on the CCA framework. Thus, also echoing the CCA, the CC(S)A emissions reduction process is based around 5-year time blocks that are in the spirit of the CCA's carbon budgeting regime[119]; however, and unlike the CCA, the Scottish framework incorporates a system of annual target values within this temporal scheme.[120] The CC(S)A's 'annual targets' obligations are sketched out over Sections 3–7, establishing a duty to set an emissions reduction target each year that aligns over time with the 2050 milestone target.[121] The Scottish Ministers are given powers to set these targets by creating Orders.[122] Advice is to be requested from the CCC that sets out views on the progress being made towards annual targets, in addition to bespoke advice of a similar nature that is to be issued by the CCC in relation to progress being made towards the interim target and the 2050 target.[123] This use of quantitative annual emissions reduction targets is an innovative feature of the CC(S)A, which differs in form somewhat to the more overt reliance on broader five-year budgets established under the CCA; this element of the CC(S)A has been characterised by environmental think tank *Ecologic* as being 'more granular' in nature.[124] A further distinct and innovate feature of the CC(S)A that differentiates the framework to some extent from the CCA concerns its

---

[118]CC(S)A, s.2(1).

[119]CC(S)A, s.4(2).

[120]CC(S)A, ss.3–4.

[121]See CC(S)A, ss.3–7.

[122]CC(S)A, s.4(1). In Scotland's subnational governance setting, these sorts of Orders are commonly viewed as a type of 'secondary' legislation, that enables the Scottish Ministers (in this case) to exercise powers provided for under primary legislation (here, the CC(S)A). For further detail on the power to create Orders under the CC(S)A, and also the power to create regulations under the Act, see CC(S)A, s.96 'Subordinate legislation'.

[123]CC(S)A, s.9(1)(a)(i)–(iii).

[124]Ecologic Institute, "Paris Compatible" Governance: Long-Term Policy Frameworks to drive Transformational Change (Ecologic Institute, 2017), p. 72.

'public engagement' elements,[125] which include a requirement placed on the Scottish Ministers to produce a *Public Engagement Strategy* and to raise general awareness about climate change amongst Scottish citizens, with the published strategy being kept under review over time and updated as appropriate.[126]

The CC(S)A's overall framing is structured around the imposition of emissions reduction targets,[127] advisory functions,[128] reporting duties,[129] duties of public bodies relating to climate change[130] and 'other climate change provisions'.[131] This latter category—'other climate change provisions'—includes amongst other things significant provision for climate change adaptation, which takes the form of producing programmes and reports in a manner that echoes to a large extent the CCA approach to adaptation.[132] These elements include the requirement that, where the UK-level Secretary of State is laying national reports before UK Parliament in relation to adaptation (under Section 56 of the CCA), the Scottish Ministers must also lay a programme before the Scottish Parliament that sets out the following:

- objectives in relation to adaptation to climate change;
- proposals and policies for meeting those objectives;
- arrangements for involving employers, trade unions and other stakeholders in meeting those objectives;
- mechanisms for ensuring public engagement in meeting those objectives;
- the period within which those proposals and policies will be introduced.[133]

The programme must also address any other risks that have been identified in the UK-level report delivered under Section 56 of the national

---

[125] See CC(S)A, s.91, 'Public engagement'.

[126] CC(S)A, s.91, 'Public engagement'.

[127] CC(S)A, Part 1.

[128] CC(S)A, Part 2.

[129] CC(S)A, Part 3.

[130] CC(S)A, Part 4.

[131] CC(S)A, Part 5.

[132] CC(S)A, 'Adaptation', ss.53–56.

[133] CC(S)A, s.53(2)(a)(i)–(v).

CCA.[134] Like the CCA, the CC(S)A incorporates carbon dioxide, methane, nitrous oxide, hydrofluorocarbons, perfluorocarbons and sulphur hexafluoride within the meaning of 'greenhouse gas'[135]; however, the Scottish have also added the additional greenhouse gas 'nitrogen trifluoride' to this cohort,[136] doing so in 2015 by issuing an Order called The Climate Change (Additional Greenhouse Gas) (Scotland) Order 2015. This Order amended the part of the CC(S)A that defines the pertinent gases. Since 2015, nitrogen trifluoride has been incorporated into UNFCCC greenhouse gas guidelines, and so it can be argued strongly that the CCA is in effect lagging behind (at the time of writing) in not updating its targeted greenhouse gas cohort to formally incorporate this additional gas.

As noted, the CC(S)A is modelled on the CCA to a significant extent. Like the CCA, it provides a broad legal architecture that enables political-legal decisions and other practical strategic governmental choices to be made and applied (via orders and regulations) as a means of realising the Act's overarching objectives. Construed in their broadest terms, those overarching objectives are to engage successfully in effective climate change mitigation and adaptation practices within Scotland. In Chapter 1 of this book, the CCA was described as something of a 'skeleton' framework that facilitates a potentially broad range of targeted and/or pragmatic political-legal action. Where these existing capacities for action and innovation are realised (within the framework's permitted parameters), this serves to put flesh on the bones of the skeleton framework and moves things towards the concrete achievement of the required long-term outcomes. Roughly the same characterisation can be applied to the CC(S)A architecture in its subnational Scottish setting.

The CC(S)A places the duty to secure mandatory emissions reductions, and its other associated primary duties (reporting, producing plans and programmes, etc.), on the 'Scottish Ministers', namely the devolved Scottish Government. This can be contrasted with the UK-level CCA, which places its primary obligations on the Secretary of State.[137] In other

---

[134] CC(S)A, s.53(2)(b).

[135] CC(S)A, s.10(1)(a)–(f).

[136] CC(S)A, s.10(1)(g). This new addition is given a baseline year of 1995; CC(S)A, s.11(2)(f). Wales has also incorporated this gas into its legislation, see Environment (Wales) Act 2016, s.37(1)(g).

[137] Specifically, the Secretary of State for Business, Energy and Industrial Strategy, as outlined in Chapter 2.

words, it can be seen that the CCA places the chief burden of account-ability on a single UK Government Minister at the head of the appro-priate government department—the Department for Business, Energy and Industrial Strategy—whereas the Scottish legislation places responsi-bility on the Scottish Ministers collectively. These alternative approaches have arisen primarily as a natural consequence of differences in the UK's national and (Scottish) substate governmental structures, that is to say, they are to be largely expected and are not unusual, innovative features of the Climate Change Acts themselves. UK Government has no corpo-rate legal identity, and legal powers/duties are commonly placed in the hands of individual Ministers. By contrast, the Scottish Government's legal powers/duties are vested in the hands of the Scottish Ministers col-lectively. On the one hand, the Scottish design seems appealing here, in that the Scottish Government is not divided along departmental lines in a way where departmental divisions might create overt barriers to inte-grated policymaking and action. As such, the Scottish model may hold positive lessons for governance. On the other hand, however, this plu-ral degree of accountability risks construing things so that no explicitly identifiable senior governmental actor is overtly accountable for fulfilling the framework's important obligations. This can be contrasted with the CCA, which foregrounds the Secretary of State as an explicitly responsi-ble party.

The Scottish framework construes the CCC as acting as a primary advisory body across the CC(S)A decarbonisation process, with its duties harmonising with the CCC obligations that are set out in the CCA.[138] It is notable, however, that the CC(S)A also enables the Scottish adminis-tration to create an internal Scotland-specific advisory body that can ful-fil these sorts of duties for Scotland instead of the CCC. Nonetheless, the CCC has fulfilled these functions to date. Most particularly, CC(S)A Section 25 enables the creation of a Scottish Committee on Climate Change, that is, a body amounting to a substate Scotland-specific ver-sion of the national CCC.[139] CC(S)A Schedule 1 sets out the form and powers of this subnational Committee, echoing somewhat the manner in which the CCA Schedules qualify the structure and composition of the

---

[138] See further Chapter 2.
[139] CC(S)A, s.25.

CCC.[140] As noted, to date the Scottish Committee on Climate Change has not been created, so that the capacity provided by the legislation to activate this type of body remains, but has not been used. Instead, the CCC has been the primary body charged with fulfilling the major advisory and reporting functions.

Section 33 of the CC(S)A requires the production of an annual report by the Scottish Government on progress towards its emissions reduction targets.[141] Where reduction obligations have not been met, the Ministers are required to 'lay a report before the Scottish Parliament, which sets out proposals and policies to compensate in future years for the excess emissions'.[142] Chapter 1 of this book has stressed that one feature of the CCA depending on one's view arguably amounts to a significant weakness concerns the fact that the framework does not contain sanctions for a failure to meet its targets. Nevertheless, it remains the case that the CCA has both engendered significant practical results and made the most indispensable legislative contribution to the UK's Low Carbon Transition, a lack of sanctions notwithstanding. Telescoping the spirit of this macro 'UK/CCA' vista down to the micro 'Scotland/CC(S)A' experience, essentially the same set of generalised observations can be drawn in the case of Scotland. Reid has emphasised that the CC(S)A fails to prescribe a specific enforcement procedure or sanctions in instances where targets are breached; however, he also notes that mandatory regular reports to the Scottish Parliament in conjunction with the advisory and reporting work provided by the CCC are designed to secure compliance.[143] Thus, while explicit sanctions are absent, they do not appear to be strictly necessary for the existence of successful action in Scotland, as illustrated by the extensive deployment of renewables and the associated decline in energy emissions within the jurisdiction, a process that the CC(S)A has contributed to importantly. Failure to achieve a CC(S)A target amounts to a breach of duty, which can be judicially reviewed in the way that CCA breaches can.[144] It is likely in such an instance that

---

[140] CCA, Schedule 1, 'The Committee on Climate Change'.

[141] CC(S)A, s.33.

[142] CC(S)A, s.36(2).

[143] C.T. Reid, 'Climate Law: Scotland', response issued to the rapporteur for the United Nations Climate Change Conference COP 15.

[144] See further the discussion of judicial review and sanctions in Chapter 1.

a court would issue 'declaratory relief', which declares a formal breach
of the law—in this case, a determination that the Scottish Ministers have
failed to achieve their statutory target—but sanctions or remedies ranging beyond this seem unlikely.[145]

The concluding portion of the CC(S)A is composed of two main
Schedules, with Schedule 1 setting out arrangements concerning the
Scottish Climate Change Committee and Schedule 2 setting out further minor changes to other items of law that are required in order to
allow the CC(S)A framework to operate correctly.[146] A third Schedule
has since been sandwiched between Schedule 1 and Schedule 2, entitled
Schedule 1A. It might have more usefully been entitled 'Schedule 3' and
placed after Schedules 1 and 2, but when it comes to form and expression environmental law does not always take the easiest route! Schedule
1A was added by the Regulatory Reform (Scotland) Act 2014, and it
relates to charges applied for the supply of carrier bags,[147] which echo at
the Scottish level the charges applied for carrier bags under the CCA at
the national level.[148]

Although the Scottish framework does follow the CCA to a significant extent, certain features here and there in addition to those already
highlighted above amount to examples of expressly Scottish innovation, or at least can be said to come with a special Scottish 'flavour'.
For example, the Scottish Parliament does not possess a great deal of
devolved energy powers in any broad sense,[149] but it does possess

---

[145] See further Reid's consideration of the Scottish experience in: C.T. Reid, 'A New Sort of Duty? The Significance of "Outcome" Duties in the Climate Change and Child Poverty Acts' 4 *Public Law* 749 (2012); C.T. Reid, 'Scotland: Constraints and Opportunities in a Devolved System' in M. Peeters, M. Stallworthy, J.C. de Larragán (eds.) *Climate Law in EU Member States: Towards National Legislation for Climate Protection* (Edward Elgar, Cheltenham, 2012).

[146] CC(S)A, Schedule 1, Schedule 2.

[147] CC(S)A, Schedule 1A, 'FIXED PENALTIES' (these letters appear as capitals in the Act). See also CC(S)A, s.88A, 'Carrier bag offences: fixed penalty notices'.

[148] See further the examination of the CCA's provisions in Chapter 2, including carrier bag charges. The Scottish equivalent of those charges is enabled at CC(S)A, s.88.

[149] The devolved energy powers across the UK's subnational jurisdictions, including Scotland, are mapped out in Muinzer and Ellis, supra, n. 67. See further G. Little, 'Energy and the Scotland Act 2016' 20(3) *Edinburgh Law Review* 394 (2016).

significant energy *efficiency* powers,[150] and the CC(S)A makes the most of this area by laying out fairly extensive energy efficiency provisions over Sections 60–74.[151] Particular duties are placed on the Scottish Ministers to promote and develop energy efficiency and renewable heat,[152] and amongst other things, the energy efficiency provisions enable the creation of 'energy efficiency discount schemes', where council tax paid on certain properties can be discounted due to improved energy efficiency standards.[153] Special duties are also placed on public bodies in Scotland to improve their contribution to the Low Carbon Transition.[154] A further feature of the CC(S)A that stands out where the framework is juxtaposed with the CCA concerns a requirement under the CC(S)A for the creation of a special 'land use strategy', intended to help land use practices in Scotland cohere with sustainability and wider decarbonisation objectives.[155] The strategy itself is intended to be a basis for practical action, and a revised updated version of the strategy is to be laid before the Scottish Parliament no later than every 5 years.[156] Another Scottish characteristic distinctly present in the legislation concerns a power to vary the permitted times in Scotland for making 'muirburn',[157] a Scottish word for 'moorburn', referring to the action of burning heather and other vegetation on a moor so that ground is cleared for new growth.[158]

In terms of litigation, although it has not yet been the subject of significant litigation in the Scottish courts, the CC(S)A has exhibited a tendency to rear its head as a background feature over the course of court

---

[150] See Muinzer and Ellis, ibid.

[151] CC(S)A, ss.60–74. Energy efficiency also surfaces as a significant theme at other important points in the framework, see, e.g., CC(S)A, s.76(1)(a).

[152] CC(S)A, s.60, s.61.

[153] CC(S)A, s.65 (substantially amending the Local Government Finance Act 1992).

[154] See CC(S)A, Part 4.

[155] CC(S)A, s.57, 'Duty to produce a land use strategy'. See also CC(S)A, ss.84–87, which enable the creation of a 'deposit and return' scheme intended to improve recycling. See further Chapter 2 on waste reduction under the CCA.

[156] CC(S)A, s.57(6).

[157] CC(S)A, s.58. The time variation is applied through amendment by the CC(S)A to the Hill Farming Act 1946.

[158] *Oxford English Dictionary (Online)*, 'moorburn, n.' Unpaginated resource.

argument or reasoning in certain instances, in particular where disputes over wind farm planning permissions are concerned.[159] This is exemplified by the opening paragraph of Lord Boyd of Duncansby's judgement in his resolution of a wind farm planning dispute in the case *Wildland Ltd and the Welbeck Estates v Scottish Ministers*,[160] which begins as follows:

> As part of its commitment to tackling climate change the Scottish Parliament passed the Climate Change (Scotland) Act 2009. That Act set out targets for reducing greenhouse gas emissions to an interim target of 42% by 2020 and an 80% target by 2050. One of the main ways of achieving these targets is the replacement of carbon emitting energy generation with renewable sourced energy. The Scottish Government has set a target of providing 30% of overall energy demand from renewable sources by 2020. A significant proportion of this will come from both offshore and onshore wind. However the development of wind power brings its own environmental challenges as turbines can have significant impacts on natural habitats, birds, landscape and scenic values. The resolution of these conflicts can raise issues of fine judgement and generate significant controversy as the number of cases involving wind farms in this court can testify.[161]

This statement captures the spirit of the manner in which the macro-architecture of the CC(S)A can stimulate and encompass the micro-issues to hand that the court is expressly concerned with, in this case involving a narrow challenge to a planning permission granted by the Scottish Ministers to a wind farm development.[162]

---

[159] The author is grateful to Sir Crispin Agnew QC (senior Scottish advocate) for drawing attention to these circumstances over the course of research for this book.

[160] *Wildland Ltd and the Welbeck Estates v Scottish Ministers* [2017] CSOH 113.

[161] Ibid., para [1].

[162] Similarly, for acknowledgement of the macro-role of the CC(S)A in relation to: the narrower issue of planning permission for housebuilding and an associated purported risk of flooding, see *Bova v Highland Council* [2013] SC 510, para [54]; an obligation to transition to renewable energy, see *Packard, Petitioner* [2011] CSOH 93, Para [19].

*

This chapter has ranged beyond national-level considerations of the CCA to address the framework's multilevel environment, exploring both its positioning and relationship to the broader international sphere and to the UK's internal subnational sphere. The following brief chapter will conclude this study by providing closing summative remarks relating to the multilevel international and subnational consideration of the CCA provided in this chapter, the analysis of the framework's background and contextual factors in Chapter 1 and the targeted explication and exploration of the content of the CCA in Chapter 2.

CHAPTER 4

Conclusions

Abstract This short closing statement draws the threads of the preceding chapters together and provides a summative conclusion.

Keywords Conclusions on the Climate Change Act 2008 · Climate and Energy Law and Policy · Climate Change Mitigation and Adaptation

*

It is by this stage clear that the UK was the first country in the world to create and apply legislation in order to set in place national long-term legally binding greenhouse gas emissions reduction targets, also incorporating into that legislation associated mechanisms intended to secure those targets in practice (most notably carbon budgeting), as well as obligations centring on the related problem of climate change adaptation. The CCA commits the UK to lowering greenhouse gas emissions on 1990 levels by at least 34% come 2020 and at least 80% come 2050. It has been seen that the regime has introduced a carbon accounting scheme broken into 5-year carbon budget cycles for the purpose, enabled the creation of special trading schemes and set the CCC in place in order to report and advise on the overall decarbonisation process. In addition to engaging with mitigation, the framework is sensitive to challenges posed

© The Author(s) 2019
T. L. Muinzer, *Climate and Energy Governance for the UK Low Carbon Transition*, https://doi.org/10.1007/978-3-319-94670-2_4

by adaptation, approaching this issue through the formation of a bespoke ASC[1] within the broader CCC, targeted adaptation reporting and the creation of practical programmes intended to counter the problem.

Set in a global context, the CCA targets are relatively ambitious and therefore should be lauded; however, the global community is not going nearly far enough to redress the scientifically projected dangers posed by anthropogenic climate change, and viewed in the light of this knowledge, it is clear that the CCA could be improved greatly if its objectives were strengthened and deepened. By 'strengthening', this should be taken to indicate most particularly the ramping up of target percentages and the tightening of other associated quantifiably measureable decarbonisation objectives generated by the framework, particularly the carbon budget thresholds. By 'deepening', this should be taken to indicate a greater socio-economic depth of reach that could be accorded to the CCA's processes, meaning in essence that the energy sector must no longer be treated as the low-hanging fruit that can bear the lion's share of emissions reductions: the UK Low Carbon Transition must be expanded and driven home through *all* socio-economic sectors as fully as possible. The CCC itself has been coming to recognise this increasingly, insofar as its more recent recommendations have been highlighting in a more emphatic way broader solutions that can be offered by sectors ranging beyond the power sector.[2] In controlling the primary levers of active policy, UK Government would be well advised to act in the spirit of this broader approach where possible, that is, the approach characterised here as the process of the 'deepening' of the CCA objectives. An ambitious, successful decarbonisation process ought to be as deepened, pervasive and totalising as possible, and within the parameters of the CCA sophisticated environmental political-legal action is required in order to realise this most effectively. This includes the augmented creation of conditions where significant research and investment capital can be streamed into both the improvement of current climate-oriented

---

[1] That is, the Adaptation Sub-Committee, as discussed in Chapter 2.

[2] See, e.g., CCC, *Sectoral Scenarios for the Fifth Carbon Budget* (CCC, 2015), where 'Decarbonising power' features as only one of 6 targeted sectoral chapters; CCC, *2017 Progress Report to Parliament—Meeting Carbon Budgets: Closing the Policy Gap* (CCC, 2017), where CCC recommendations to UK Government span significantly beyond the power sector in this ninth annual assessment of UK progress in meeting carbon budgets under the CCA.

environmental technologies and the development of future technologies, not least in the area of renewables.

In terms of practical performance, the UK has generally been successful in adhering to its CCA obligations to date and in achieving its practical targets. If one views the glass as half full, this is cause for celebration; however, if one views the glass as half empty, one must ask why the UK has not set more ambitious targets. Climate mitigation is a great global urgency, and so, given that the UK has roughly hit its interim 2020 emissions reduction target already, well in advance of the target year, and in relatively comfortable fashion, surely it is time to go further and ramp the decarbonisation trajectory up substantially (this would reflect the substate approach being adopted by Scotland at the present time[3]). This position is supported by conventional environmental ethics, insofar as an ethical imperative exists to safeguard our planet adequately for present and future generations. It is also supported by cruder but valid economic logic, given that a substantial international market in the 'green dollar' now exists, and the UK has an opportunity to improve its economic position here, where it is already a significant player in renewables and carbon markets, etc. Moreover, this is to say nothing of the costs to the national economy that can be saved overall in the area of *adaptation* by acting robustly here and now to steadily manage and increase resilience to the impacts of present and future climate extremes. Further, and as a separate point in its own right, while it is fair to say that the UK's performance under the CCA as it currently stands has been reasonably successful up to the present time, it is notable that one is dealing here with a *long-term* framework where forecasting far into the future becomes an at best highly speculative task. Thus, even on the narrow point of the future of electricity in the context of the CCA, Helm has noted that: 'The carbon budgets are already defined until 2032. Parliament has approved them all, and 15 years is a long time in the electricity sector in terms of predictability. Beyond 2032, any detailed forecasts of costs are likely to turn out wrong, and perhaps by orders of magnitude'.[4] This observation catches the spirit of the extent to which one cannot merely assume that current successes can be projected forward with some degree of casual reliability, for one is dealing with a highly uncertain future.

---

[3] See Chapter 3.

[4] D. Helm, *Cost of Energy Review* (BEIS, 2017), p. 13.

It has been seen in Chapter 1 that the CCA emerged out of a very distinct set of circumstances, and that some degree of meaningful cross-party political support and a significant level of public pressure played no small part in its genesis. The pioneering framework that resulted was not without its technical problems, which remain embedded within it at the present time. Perhaps most significantly, these problems include insecurity around the extent to which key legally binding duties under the framework can actually be meaningfully enforced. Certainly, no explicit sanctions are made available under the terms of the CCA itself that can apply where, say, the 2050 80% reduction target is missed. The extent to which the CCA has adequately accounted for the UK's subnational governance environment has also been questioned, as highlighted in Chapter 1, and a sense of the character of the UK's substate arena has been fleshed out in Chapter 3, where it has been seen that UK devolution provides capacity for meaningful political-legal action in the sphere of substate climate and energy governance. Any state seeking to apply a CCA-style model to the climate challenge should be advised to carefully factor in consideration of what are likely to be complex multitiered capacities and constraints that may arise where a blanket national-level regime is interconnected with a state's particular subnational governance environment. In the case of the UK, Chapter 3 has outlined and explored how progressive climate governance on the part of the Scottish administration below the state level has served to significantly bolster the UK decarbonisation drive. It has also been seen in that chapter that the Northern Irish administration has a great practical capacity for action due to its devolved powers, but the jurisdiction is occupying the position of the UK's climate laggard. Northern Ireland is small enough for this circumstance to put only a limited dent in the UK's aggregate Low Carbon Transition process, but if Scotland had proven to be similarly recalcitrant to date, then the combined socio-economic power of Northern Ireland and Scotland, working in conjunction with their significant devolved capacities to exhibit resistance to national-level drivers, could have potentially derailed the UK's overall decarbonisation drive. This situation could have been worse still if a recalcitrant Welsh devolved government had emerged and endeavoured to set its shoulder to this type of wheel as well.

It is clear from the multilevel considerations in Chapter 3 that, just as the UK has substantial lessons to learn from the international community, the international community has something to learn from the CCA, and it is little wonder that the UK's framework has exerted some notable degree of international influence to date. It is also clear from the discussion in Chapter 3 that the form and content of the CCA itself have been informed by international developments, and as a related but distinct point, that the CCA has been designed to harmonise to a significant extent with certain aspects of progressive international action (e.g. through standardisation of targeted greenhouse gases, definitional and reporting norms, certain key dates such as 2020 and an ability to interact with international trading schemes under the CCA trading scheme capacities). This inbuilt sensitivity to broader global developments should be recommended as an example of good practice to any state seeking to craft and apply similarly robust and progressive climate and energy mitigation and adaptation legislation.

Chapter 2 has outlined and explored the substance, form and content of the CCA itself. While the CCA amounts to a sophisticated, substantial regime, it is also the case that alternative approaches might have been employed in order to secure the sorts of outcomes that the CCA is striving for. For example, a regime that foregrounds extensive and targeted carbon taxing could have been constructed instead, or indeed a regime that is less heavily target-oriented and that puts a greater onus on markets.[5] In closing, it may be fair to say that the answer to the question as to what the most fitting state-level climate and energy governance architecture should actually 'look like' is perhaps an open one; however, any thorough answer must be informed by the knowledge and experience gained from observation of the construction, content, successes and failures of the UK's important and pioneering CCA.

[5] See Fankhauser, Averchenkova and Finnegan's comment on 'Policy design', p. 4 of their report at supra, n. 52.

Bibliography

Scholarly Commentary

S. Afionis, *The European Union in International Climate Change Negotiations* (Routledge, Abingdon, 2017).

J. Andrews, N. Jelley, *Energy Science* (2nd edition, Oxford University Press, Oxford, 2013).

NRF. Al-Rodhan (ed.) *Policy Briefs on the Transnational Aspects of Security and Stability* (LIT Verlag, Munster, 2007).

G. Anthony, *Judicial Review in Northern Ireland* (Hart, Oxford, 2014).

I. Backer, O. Fauchald, C. Voigt (eds.) *Pro Natura* (Universitetsforlaget, Oslo, 2012).

H. Baer, M. Singer, *The Anthropology of Climate Change: An Integrated Critical Perspective* (Routledge, Abingdon, 2014).

L. Benjamin, 'The Responsibilities of Carbon Major Companies: Are They (and Is the Law) Doing Enough?' (5)2 *Transnational Environmental Law* 353 (2015).

M.H. Benson, 'Regional Initiatives: Scaling the Climate Response and Responding to Conceptions of Scale' 100(4) *Annals of the Association of American Geographers* 1025 (2010).

D. Bodansky, 'The Legal Character of the Paris Agreement' 25(2) *Review of European Comparative and International Environmental Law* 142 (2016).

Lord Bourne of Aberystwyth, 'The Paris Agreement Proves That the Transition to a Climate-Neutral and Climate-Resilient World Is Happening', published speech transcript to the UN, delivered 22 April 2016, published 25 April 2016 (UK Government, 2016).

© The Editor(s) (if applicable) and The Author(s) 2019

T. L. Muinzer, *Climate and Energy Governance for the UK Low Carbon Transition*, https://doi.org/10.1007/978-3-319-94670-2

G. Boyle, *Renewable Energy: Power for a Sustainable Future* (Oxford University Press, Oxford, 2012).

G. Bridge, et al. 'Geographies of Energy Transition: Space, Place and the Low Carbon Economy' 53 *Energy Policy* 331 (2013).

C. Callaghan, 'What Is a "Target Duty"?' 5(3) *Judicial Review* 186 (2000).

H.V. Campbell, 'A Rising Tide: Wave Energy in the United States and Scotland' 2(2) *Sea Grant Law and Policy Journal* 29 (Winter 2009/2010).

N. Carter, 'Combating Climate Change in the UK: Challenges and Obstacles' 79(2) *The Political Quarterly* 194 (2008).

D. Castelvecchi, 'New Brexit Government Spells Shake-Up for Science: Theresa May Promotes a Former Science Minister and Abolishes Climate-Change Department' 535(7612) *Nature* 331 (2016).

N. Chomsky, 'Global Warming and the Common Good', Talk Delivered at East Stroudsburg University, 7 February 2013. Transcribed on the *Reading Chomsky* Website: http://readingchomsky.blogspot.co.uk/2013/04/normal-0-0-2-false-false-false-en-us-ja.html.

J. Church, 'Enforcing the Climate Change Act' 4(1) *UCL Journal of Law and Jurisprudence* 109 (2015).

R. Cowell, 'Decentralising Energy Governance? Wales, Devolution and the Politics of Energy Infrastructure Decision-Making' 35(7) *Environment and Planning C: Politics and Space* 1242 (2017).

M. Elliott, R. Thomas, *Public Law* (2nd edition, Oxford University Press, Oxford, 2014).

G. Ellis, R. Cowell, F. Sherry-Brennan, P. Strachan, D. Toke, 'Planning, Energy and Devolution in the UK' 84(3) *Town Planning Review* 397 (2013).

M. Faure, M. Peeters (eds.) *Climate Change and European Emissions Trading* (Edward Elgar, Cheltenham, 2008).

D. Feldman, 'Legislation Which Bears No Law' 37(3) *Statute Law Review* 212 (2016).

E. Fisher, E. Scotford, E. Barritt, 'The Legally Disruptive Nature of Climate Change' 2(80) *Modern Law Review* 173 (2017).

M. Fordham, *Judicial Review Handbook* (6th edition, Hart, Oxford, 2012).

R. Fouquet (ed.) *Handbook on Energy and Climate Change* (Edward Elgar, Cheltenham, 2013).

A. Frank, 'Paris Climate Agreement: Success or Failure?' *Cosmos & Culture—NPR Blog* (published electronically), 12 January 2016.

M.J. Goodwin, O. Heath, 'The 2016 Referendum, Brexit and the Left Behind: An Aggregate-Level Analysis of the Result' 87(3) *The Political Quarterly* 323 (2016).

M. Grubb, 'The Economics of the Kyoto Protocol' 4(3) *World Economics* 143 (2003).

K. Hill, *The UK Climate Change Act 2008—Lessons for National Climate Laws* (ClientEarth, London, 2009).

E.J. Hollo, K. Kulovesi, M. Mehling (eds.) *Climate Change and the Law* (Springer, Dordrecht, 2012).

J. Hovi, D.F. Sprinz, G. Bang, 'Why the United States Did Not Become a Party to the Kyoto Protocol: German, Norwegian and US Perspectives' 18(1) *European Journal of International Relations* 129 (2010).

T. Jackson, W. Lynch, 'Public Sector Responses to Climate Change: Evaluating the Role of Scottish Local Government in Implementing the Climate Change (Scotland) Act 2009' 8/9 *Commonwealth Journal of Local Governance* 112 (2011).

M. Laffin, A. Thomas, 'Designing the National Assembly for Wales' 53(3) *Parliamentary Affairs* 557 (2000).

D. Lambert, 'The Government of Wales Act—An Act for Laws to be Ministered in Wales in Like Form as It Is in This Realm?' 30 *Cambrian Law Review* 60 (1999).

J. Lang, 'Zero Time: NZ's Zero Carbon Act', *E-nvironmentalist* (published electronically, unpaginated), 18 September 2017.

R. Leal-Arcas, J. Wouters (eds.) *Research Handbook on EU Energy Law and Policy* (Edward Elgar, Cheltenham, 2017).

G. Leydier, A. Martin (eds.) *Environmental Issues in Political Discourse in Britain and Ireland* (Cambridge Scholars, Newcastle, 2013).

G. Little, 'Energy and the Scotland Act 2016' 20(3) *Edinburgh Law Review* 394 (2016).

M. Lockwood, 'The Political Sustainability of Climate Policy: The Case of the UK Climate Change Act' 23(5) *Global Environmental Change* 1339 (2013).

M. Loughlin, *The British Constitution: A Very Short Introduction* (Oxford University Press, Oxford, 2013).

A. McHarg, 'Climate Change Constitutionalism? Lessons from the United Kingdom' 2(4) *Climate Law* 469 (2011).

D. McKittrick, D. McVea, *Making Sense of the Troubles* (New Amsterdam Books, Chicago, 2002).

P. McMaster, 'Climate Change—Statutory Duty or Pious Hope?' 20(1) *Journal of Environmental Law* 115 (2008).

R. Macrory, *Regulation, Enforcement and Governance in Environmental Law* (2nd edition, Hart, London, 2014).

C. Mitchell, J. Watson, J. Whiting (eds.) *New Challenges in Energy Security: The UK in a Multipolar World* (Palgrave Macmillan, Basingstoke, 2013).

T.L. Muinzer, 'Is the Climate Change Act 2008 a "Constitutional Statute"?' *European Public Law* (2018, forthcoming).

T.L. Muinzer, G. Ellis, 'Subnational Governance for the Low Carbon Energy Transition: Mapping the UK's "Energy Constitution"' 35(7) *Environment and Planning C: Politics and Space* 1176 (2017).

T.L. Muinzer, '"To PV or Not to PV": An Analysis of the High Court's Recent Treatment of Solar Energy' 17(2) *Environmental Law Review* 128 (2015).

T.L. Muinzer, *The UK's Energy Decarbonisation Process and the Challenges of Devolution* (PhD research thesis), Queen's University Belfast, 1 October 2011–1 July 2015.

T.L. Muinzer, 'An Evaluation of the Implications of EU Climate and Energy Governance for the UK in light of Brexit' 23(2) *European Journal of Current Legal Issues* (2017).

T.L. Muinzer, 'Does the Climate Change Act 2008 Adequately Account for the UK's Devolved Jurisdictions?' 25(3) *European Energy and Environmental Law Review* 87 (2016).

T.L. Muinzer, 'Incendiary Developments: Northern Ireland's Renewable Heat Incentive, and the Collapse of the Devolved Government' (99) *UKELA E-Law* 18 (2017), March/April.

T.L. Muinzer, 'Warming UP: Northern Ireland's Developing Response to Climate Change in the Context of UK Devolution' (96) (September/October) *UKELA E-Law* 19 (2016).

M. Navarro, 'A Substantial Body of Different Welsh Law: A Consideration of Welsh Subordinate Legislation' 33(2) *Statute Law Review* 163 (2012).

S. Oberthur, H.E. Ott, *The Kyoto Protocol: International Climate Policy for the 21st Century* (Springer, Dordrecht, 1999).

S. Oberthur, C.R. Kelly, 'EU Leadership in International Climate Policy: Achievements and Challenges' 43(3) *The International Spectator* 35 (2008).

R. Owen, 'Government of Wales Act 2006' 42(1) *The Law Teacher* 103 (2008).

R. Owen, 'Should Wales Separate from England's Legal System?', *The Conversation* (published electronically, unpaginated), 12 April 2016.

N. Parpworth, *Constitutional and Administrative Law* (8th edition, Oxford University Press, Oxford, 2014).

M.R. Pasimeni, et al. 'Scales, Strategies and Actions for Effective Energy Planning: A Review' 65 *Energy Policy* 165 (2014).

M. Peeters, M. Stallworthy, J.C. de Larragán (eds.), *Climate Law in EU Member States: Towards National Legislation for Climate Protection* (Edward Elgar, Cheltenham, 2012).

R.A. Pielke, 'The British Climate Change Act: A Critical Evaluation and Proposed Alternative Approach' 4(2) *Environmental Research Letters* 1 (2009).

C.T. Reid, 'A New Sort of Duty? The Significance of "Outcome" Duties in the Climate Change and Child Poverty Acts' 4 *Public Law* 749 (2012).

C.T. Reid, 'Climate Law: Scotland', Response issued to the rapporteur for the United Nations Climate Change Conference COP 15.

C.T. Reid, 'Climate Change Law in Scotland' Ympäristö-Juridiikka Miljöjuridik (1) *Finnish Environmental Law Review* 18 (2012).

G. Robinson, 'Stemming the Rising Tide: Developing Approaches for UK Law and Policy to Combat Climate Change' 2(1) *Birmingham Student Law Review* 27 (2017).

J. Robinson, J. Barton, C. Dodwell, M. Heydon, L. Milton, *Climate Change Law: Emissions Trading in the EU and the UK* (Cameron May, London, 2007).

P. Sands, J. Peel, A. Fabra, R. MacKenzie, *Principles of International Environmental Law* (4th edition, Cambridge University Press, Cambridge, 2018).

S. Schiele, *Evolution of International Environmental Regimes: The Case of Climate Change* (Cambridge University Press, Cambridge, 2014).

R.A. Schultz, *Technology Versus Ecology: Human Superiority and the Ongoing Conflict with Nature* (IGI, Hershey, 2013).

A.J. Simcock, 'One and Many—The Office of Secretary of State' 70(4) *Public Administration* 535 (1992).

S. Smith, T. Dunne, A. Hadfield (eds.) *Foreign Policy: Theories, Actors, Cases* (3rd edition, Oxford University Press, Oxford, 2016).

B.K. Sovacool, M.A. Brown 'Scaling the Policy Response to Climate Change' 27(4) *Policy and Society* 317 (2009).

M. Stallworthy, 'Legislating Against Climate Change: A UK Perspective on a Sisyphean Challenge' 72(3) *Modern Law Review* 412 (2009).

M. Stallworthy, 'New Forms of Carbon Accounting: the Significance of a Climate Change Act for economic Activity in the UK' *International Company & Commercial Law Review* 331 (2007).

J. von Stein, 'The International Law and Politics of Climate Change: Ratification of the United Nations Framework Convention and the Kyoto Protocol' 52(2) *Journal of Conflict Resolution* 243 (2008).

N. Stern, 'What Is the Economics of Climate Change?' 7(2) *World Economics* 1 (2006).

D. Toke, F. Sherry-Brennan, R. Cowell, G. Ellis, P. Strachan, 'Scotland, Renewable Energy and the Independence Debate: Will Head or Heart Rule the Roost?' 84(1) *Political Quarterly* 61 (January–March) (2013).

H. Townsend, 'The Climate Change Act 2008: Something to Be Proud of After All?' 7(8) *Journal of Planning and Environmental Law* 842 (2009).

H. Townsend, 'Climate Change Act 2008: Will It Do the Trick?' 11(2) *Environmental Law Review* 116 (2009).

A. Trench, 'The Government of Wales Act 2006: The Next Steps on Devolution for Wales' *Public Law* 687 (2006).

S. Turner, 'Committing to Effective Climate Governance in Northern Ireland: A Defining Test of Devolution' 25(2) *Journal of Environmental Law* 203 (2013).

S. Turner, 'Northern Ireland's Consent to the Climate Change Act 2008: Symbol or Illusion?' 25(1) *Journal of Environmental Law* 63 (2013).

C. Turpin, A. Tomkins, *British Government and the Constitution* (6th edition, Cambridge University Press, Cambridge, 2007).

B. Ward, 'How Will Brexit Affect Climate Change Policy?' *News & Commentaries* (published electronically, unpaginated; Grantham Research Institute on Climate Change and the Environment), 30 June 2016.

J. Wouters, H. Bruyninckx, S. Basu, S. Schunz (eds.) *The European Union and Multilateral Governance: Assessing EU Participation in United Nations Human Rights and Environmental Fora* (Palgrave Macmillan, Basingstoke, 2012).

(unattributed) 'Energy Legislation: the Climate Change Act 2008' (legislative comment) *Environmental Law Monthly* 1 (2008).

Reports & Other

Reports

J. Aldy, et al., *Beyond Kyoto: Advancing the International Effort Against Climate Change* (Pew Center on Global Climate Change, 2003).

CCC, *2017 Progress Report to Parliament—Meeting Carbon Budgets: Closing the Policy Gap* (CCC, 2017).

CCC, *The Appropriateness of a Northern Ireland Climate Change Act* (CCC, 2011).

CCC, *The Appropriateness of a Northern Ireland Climate Change Act—December 2015 Update* (CCC, 2015).

CCC, *Reducing Emissions in Scotland: 2017 Progress Report to Parliament* (CCC, 2017).

CCC, *Sectoral Scenarios for the Fifth Carbon Budget* (CCC, 2015).

Council of the European Union, 'Presidency Conclusions', Brussels European Council, 8/9 March 2007, 7224/1/07 CONCL 1.

CUNCR, *Is There Global Climate Justice?—Outcome Document* (CUNCR, 2017).

DECC, *The Carbon Plan* (HM Government, 2011).

DECC, *Impact Assessment for the Level of the Fifth Carbon Budget* (HM Government, 2016).

DECC, *UK Low Carbon Transition Plan: National Strategy for Climate and Energy* (HM Government, 2009).

DECC, *UK Progress Towards GHG Emissions Reduction Targets: Statistical Release* (HM Government, 2015).

DEFRA, *Agriculture in the United Kingdom 2016* (HM Government, 2017).

DEFRA, *Climate Change: The UK Programme 2006* (HM Government, 2006).

DEFRA, *UK Climate Change Risk Assessment: Government Report* (HM Government, 2012).

DEFRA, *Taking Forward the UK Climate Change Bill: The Government Response to Pre-legislative Scrutiny and Public Consultation* (HM Government, 2007).

DETR, *Climate Change: The UK Programme* (HM Government, 2000).

DfT, *Guidance: Renewable Transport Fuels Obligation* (HM Government, 2012).

DfT, *RTFO Guidance Part One, Process Guidance* (HM Government, 2017).

DOE, *Synopsis of Responses to the Department's Pre-consultation Seeking Views on the Need for a Northern Ireland Climate Change Bill* (DOE, 2013).

DTI, *Meeting the Challenge: A White Paper on Energy* (HM Government, 2007).

DTI, *Our Energy Future—Creating a Low Carbon Economy* (HM Government, 2003).

Ecologic Institute, *"Paris Compatible" Governance: Long-Term Policy Frameworks to Drive Transformational Change* (Ecologic Institute, 2017).

European Commission, *Energy Roadmap 2050* (COM/2011/885).

European Commission, *Limiting Global Climate Change to 2 Degrees Celsius: The Way Ahead for 2020 and Beyond*, COM (2007) 2 final.

European Commission, *A Policy Framework for Climate and Energy in the Period from 2020 to 2030*, COM(2014) 15 final/2.

European Commission, *A Roadmap for Moving to a Competitive Low Carbon Economy in 2050* (COM/2011/112).

S. Fankhauser, A. Averchenkova, J. Finnegan, *10 Years of the UK Climate Change Act* (Grantham Research Institute & London School of Economics, 2018).

G. Ellis, R. Cowell, F. Sherry-Brennan, P. Strachan, D.Toke, *Delivering Renewable Energy Under Devolution: Initial Findings Summary Report* (DREUD, 2013).

D. Helm, *Cost of Energy Review* (BEIS, 2017).

IDLO, *The New General Law on Climate Change in Mexico: Leading National Action to Transition to a Green Economy* (IDLO, 2012).

IPCC, *Climate Change 2013: The Physical Science Basis. Contribution of Working Group I to the Fifth Assessment Report* (Cambridge University Press, 2014).

Joint Committee on the Draft Climate Change Bill, *Oral and Written Evidence (Second Report); Draft Climate Change Bill* (2006–2007, HL 170-II, HC 542-II).

M. Nachmany, S. Fankhauser, et al., *The 2015 Global Climate Legislation Study: A Review of Climate Change Legislation in 99 Countries, Summary for Policy-Makers* (Grantham Research Institute, GLOBE, IPU, 2015).

National Records of Scotland, *Mid-2016 Population Estimates Scotland* (Scottish Government, 2017).

Northern Ireland Executive, *Programme for Government 2011–15* (NIE, 2011).

Office for National Statistics, *Overview of the UK Population: July 2017* (ONS, 2017).

Papua New Guinea Office of Climate Change and Development, *Papua New Guinea's Commitment to Act on Climate Change* (Papua New Guinea Government, 2010).

RTPINI, *Pre-consultation Seeking Views on the Need for a Northern Ireland Climate Change Bill: A Response by the Royal Town Planning Institute Northern Ireland* (RTPINI, 2013).

Scottish Government, *A Nation with Ambition: The Government's Programme for Scotland 2017–18* (Scottish Government, 2017).

Scottish Government, *Proposals for a New Climate Change Bill: Strategic Environmental Assessment Environmental Report* (Scottish Government, 2017).

Scottish Government, *Scottish Greenhouse Gas Emissions 2015* (Scottish Government, 2015).

N. Stern, et al., *Stern Review: The Economics of Climate Change* (HM Government, 2006).

Welsh Assembly Government, *Climate Change Strategy for Wales* (WAG, 2010).

Welsh Assembly Government, *Energy Wales: A Low Carbon Transition* (WAG, 2012).

T. Weeks (for the New Zealand Productivity Commission), *Examining the UK Climate Change Act 2008* (New Zealand Productivity Commission, 2017).

Newspapers

'Analysis—Roger Harrabin, Environment Analyst', Set Within the News Item 'Government Axes Climate Department' by P. Rincon, *BBC News* (online), 14 July 2016.

'Environment Minister Rules Out Separate Climate Change Laws', *Belfast Telegraph* (UK newspaper), 5 December 2016.

M. McKimm, 'Northern Ireland "Should Do More over Carbon Emissions"' *BBC News* (online), 8 November 2011.

I. Johnson, 'Climate Change Department Closed by Theresa May in "Plain Stupid" and "Deeply Worrying" Move', *Independent* (UK newspaper), 14 July 2016.

Scottish Government (press release), 'Leading the Way on Climate Change', 10 November 2017.

Websites

The 'Big Ask' campaign's homepage, hosted at the *Friends of the Earth Europe* website: http://www.foeeurope.org/the-big-ask.

The Big Ask—Questions and Answers, at the *Friends of the Earth Europe* website: http://www.foeeurope.org/node/670.

Glossary on the Official UK Parliament website: http://www.parliament.uk/site-information/glossary/orders-in-council/.
Oxford English Dictionary (online): http://www.oed.com.
Scottish Government, *Energy in Scotland: Get the Facts*, hosted at the Scottish Government's online database): http://www.gov.scot/Topics/Business-Industry/Energy/Facts.
The Scottish Government's *Greener Scotland* website: http://www.greenerscotland.org.

OTHER

Early Day Motion (EDM), No. 178, 24 April 2005.
HM Government, *Draft Climate Change Bill*, Cm 7040, March 2007.
Letter from Adair Turner, Chair of the CCC, to the Secretary of State, headed *Interim Advice by the Committee on Climate Change*, 7 October 2008.

LEGISLATION

Adoption of the Paris Agreement, FCCC/CP/2015/L.9/Rev.1.
Bali Action Plan, Decision 1/CP.13.
Bill of Rights 1688.
Clean Neighbourhoods and Environment Act 2005.
Climate Change Act 2008.
Climate Change Act 2008 (2020 Target, Credit Limit and Definitions) Order 2009.
Climate Change (Management) Act 2015.
Climate Change (Scotland) Act 2009.
Climate Change and Sustainable Energy Act 2006.
Constitution of France.
Constitution of Ireland—*Bunreacht na hEireann*.
Copenhagen Accord of 18 December 2009, Decision 2/CP.15.
Council Decision 406/2009/EC [2009] OJ L 140/136 ('Effort Sharing Decision').
Council Decision 1639/2006/EC [2006] OJ L 310/15.
Council Directive 2002/91/EC [2002] OJ L1/65.
Council Directive 2006/32/EC [2006] OJ L114/64.
Council Directive 2009/28/EC [2009] OJ L140/16 ('Renewables Directive').
Council Directive 2009/29/EC [2009] OJ L140/63 ('ETS Directive').
Council Directive 2009/125/EC [2009] OJ L285/10.
Council Directive 2009/231/EC [2009] OJ L140/114 ('CCS Directive').

Council Directive 2010/30/EC [2010] OJ 153/1.
Council Directive 2010/31/EU [2010] OJ L153/13.
Council Directive 2012/27/EU [2012] OJ L315/1 ('Energy Efficiency Directive').
Energy Act 2004.
Energy Act 2011.
Energy Act 2016.
Environmental Permitting (England and Wales) Regulations 2007.
Environmental Permitting (England and Wales) Regulations 2010.
Environmental Permitting (England and Wales) Regulations 2016.
Environmental Protection Act 1990.
Environment (Wales) Act 2016.
European Convention on Human Rights.
Government of Wales Act 1998.
Government of Wales Act 2006.
Hill Farming Act 1946.
Human Rights Act 1998.
Infrastructure Act 2015.
Kyoto Protocol (Amendment) Decision 1/CMP.8.
Kyoto Protocol to the United Nations Framework Convention on Climate Change.
Local Government Finance Act 1992.
Northern Ireland Act 1998.
Pollution Prevention and Control Act 1999.
Renewable Transport Fuel Obligations Order 2007.
Scotland Act 1998.
Single Use Carrier Bags Charge (Scotland) Regulations 2014.
United Nations Framework Convention on Climate Change.
Wales Act 2017.
Warm Homes and Energy Conservation Act 2000.
Waste Management Licensing Regulations 1994.
Well-Being of Future Generations (Wales) Act 2015.

CASES

Bova v Highland Council [2013] SC 510.
H. v Lord Advocate [2012] UKSC 308.
In the Matter of an Application By JR 47 for Judicial Review [2013] NIQB 7.
Landscape Guardians Inc v Minister for Planning [2007] NSWLEC 59
Packard, Petitioner [2011] CSOH 93.
Preston New Road Action Group v Secretary of State for Communities and Local Government [2018] EWCA Civ 9.

The Queen on the Application of London Borough of Hillingdon & Ors v Secretary of State for Transport v Transport for London [2010] EWHC 626 (Admin).

R v London Borough of Islington ex p. Rixon [1997] ELR 66.

R (G) v Barnett LBC [2004] 2 AC 208.

R (on the application of Drax Power Ltd) v HM Treasury [2016] EWHC 228 (Admin).

R (on the application of Friends of the Earth) v Secretary of State for Energy & Climate Change [2009] EWCA Civ 810.

R (on the application of Griffin) v Newham LBC Divisional Court [2011] EWHC 53 (Admin).

R (on the application of People & Planet) v HM Treasury [2009] EWHC 3020 (Admin).

R (Plan B Earth and Others) v. Secretary of State for Business Energy and Industrial Strategy (Defendant) and the Committee on Climate Change (Interested Party) [2018] EWHC 1892 (Admin).

Solar Century Holdings Limited & Others v Secretary of State for Energy & Climate Change [2014] EWHC 3677 (Admin).

Thoburn v Sunderland City Council [2003] QB 151.

Urgenda Foundation v Netherlands (24 June 2015) ECLI:NL:RBDHA:2015:7196.

Wildland Ltd and the Welbeck Estates v Scottish Ministers [2017] CSOH 113.

INDEX

© The Editor(s) (if applicable) and The Author(s) 2019

T. L. Muinzer, *Climate and Energy Governance for the UK Low Carbon Transition*, https://doi.org/10.1007/978-3-319-94670-2